T0205618

Springer Theses

Recognizing Outstanding Ph.D. Research

Aims and Scope

The series "Springer Theses" brings together a selection of the very best Ph.D. theses from around the world and across the physical sciences. Nominated and endorsed by two recognized specialists, each published volume has been selected for its scientific excellence and the high impact of its contents for the pertinent field of research. For greater accessibility to non-specialists, the published versions include an extended introduction, as well as a foreword by the student's supervisor explaining the special relevance of the work for the field. As a whole, the series will provide a valuable resource both for newcomers to the research fields described, and for other scientists seeking detailed background information on special questions. Finally, it provides an accredited documentation of the valuable contributions made by today's younger generation of scientists.

Theses are accepted into the series by invited nomination only and must fulfill all of the following criteria

- They must be written in good English.
- The topic should fall within the confines of Chemistry, Physics, Earth Sciences, Engineering and related interdisciplinary fields such as Materials, Nanoscience, Chemical Engineering, Complex Systems and Biophysics.
- The work reported in the thesis must represent a significant scientific advance.
- If the thesis includes previously published material, permission to reproduce this must be gained from the respective copyright holder.
- They must have been examined and passed during the 12 months prior to nomination.
- Each thesis should include a foreword by the supervisor outlining the significance of its content.
- The theses should have a clearly defined structure including an introduction accessible to scientists not expert in that particular field.

More information about this series at http://www.springer.com/series/8790

Harley Scammell

Interplay of Quantum and Statistical Fluctuations in Critical Quantum Matter

Doctoral Thesis accepted by
the University of New South Wales, Sydney, Australia

Author
Dr. Harley Scammell
ARC Centre of Excellence in Future Low-Energy Electronics Technologies
University of New South Wales
Sydney, NSW, Australia

Supervisor
Prof. Oleg Sushkov
University of New South Wales
Sydney, NSW, Australia

ISSN 2190-5053 ISSN 2190-5061 (electronic)
Springer Theses
ISBN 978-3-030-07369-5 ISBN 978-3-319-97532-0 (eBook)
https://doi.org/10.1007/978-3-319-97532-0

This Springer imprint is published by the registered company Springer Nature Switzerland AG
The registered company address is: Gewerbestrasse 11, 6330 Cham, Switzerland

The work teaches you how to do it.

—A fortune cookie

To mum and dad

Supervisor's Foreword

The general theme of Harley Scammell's Ph.D. thesis is the application of quantum field theories to quantum magnets. Quantum magnets are strongly correlated quantum systems, and one invariably requires novel techniques and approaches in order to understand the behavior of these systems. Specifically, this thesis considers quantum antiferromagnets, magnon Bose-condensates, and systems exhibiting deconfined quantum criticality. The main technical achievement of this thesis is the incorporation of both quantum and statistical fluctuations into a quantum field theoretic treatment of critical phenomena. This offers a significant new insight into an abundance of problems, puts them in a much more general context, and provides an unmatched power in analyzing experimental and numerical data as well as predicting new effects.

Harley begins the thesis with his exciting results on quantum antiferromagnets. This work was strongly motivated by experiments with 3+1-dimensional quantum critical magnet $TlCuCl_3$. While the theory of 2+1 quantum criticality is very well developed, especially in relation to cuprate superconductors, the 3+1 case has not been studied in as great detail. Harley's first contribution to this field concerned the logarithmic renormalization group (RG) running of the coupling constant in the vicinity of quantum critical point. Of course the running itself was known in theory for decades; however, the theory was developed only for zero temperature quantum phase transitions (equivalent to classical 4-dimensional theory). At the same time, the most important $TlCuCl_3$ data are taken at nonzero temperatures. Harley has developed techniques to handle the problem and performed RG + temperature calculations. Agreement of the theory with experiment is excellent. Analysis of experimental data has unambiguously pinned down the RG running of the coupling constant.

Harley goes on to consider the highly interesting and yet mathematically dubious regime known as dimensional crossover, which is understood to occur in the vicinity of the nonzero temperature phase transition in 3+1 quantum antiferromagnets. In this regime, it is understood that the dimension changes from 3+1 to 3, and that conventional perturbative quantum field theoretic approaches breakdown (typically due to infrared divergences). Harley approaches the problem from two distinct starting points:

(1) *Dynamic observables*. First, Harley develops a novel technique for the summation of infrared divergent diagrams in the vicinity of the Neel temperature, i.e., dimensional crossover. Using the developed technique, Harley calculates the lifetimes of magnetic excitations in 3+1 quantum critical magnets. Comparison between theory and experimental results on $TlCuCl_3$ shows excellent agreement.

(2) *Static observables*. Next, Harley demonstrates a way to reorganize perturbation theory, which allows for a continuous description of both the 3+1-dimensional regime and the 3-dimensional regime. Moreover, to an excellent approximation, it reproduces the known perturbative RG results which can be obtained by assuming either 3+1- or 3-dimensional field theory separately.

Another significant set of results relate to magnon Bose-condensates. First, Harley develops a consistent theoretical framework that resolves a long-standing inconsistency in the theory of Bose-condensation. It has been known for a long time that the Popov diagrammatic technique leads to a discontinuous Bose-condensation transition; even so many approaches to magnon Bose-condensation in the modern literature still rely on this technique. Harley provides a resolution to the problem using quantum field theory. Using the developed approach as a starting point, Harley obtains several new results:

(i) Two new universality classes are uncovered within the magnon Bose-condensation phase diagram. (ii) A long-standing issue with a known critical index is resolved. The resolution in part relies on the introduction of the logarithmic running coupling constant. (iii) The prediction of an ultra-narrow Higgs resonance in magnon Bose-condensates.

A final major result relates to the exotic scenario of deconfined quantum criticality. Within this framework, the thesis predicts Bose-condensation of particles with half-integer spin, the first ever suggestion. A smoking gun criterion to test for this exotic condensate is established.

Sydney, Australia
July 2018

Prof. Oleg Sushkov

Abstract

This thesis is an exploration of critical phenomena in highly correlated quantum matter. Specifically, we consider quantum antiferromagnets, magnon Bose-condensates, and systems exhibiting deconfined quantum criticality. Within these systems, the critical phenomena of interest are the *static* properties—quantum and classical critical points, ground-state symmetries, the order parameter, critical indices, and universality classes—as well as the *dynamic* properties—excitation interactions, lifetimes, and energies.

This thesis may be partitioned according to the system under investigation. In the first part of this thesis, we consider quantum antiferromagnets. We derive finite temperature properties of a quantum field theory and use the results to analyze experimental data. The analysis provides the first identification of asymptotic freedom of magnetic excitations. We further verify our findings against high-precision numerical quantum Monte Carlo data. Next, we consider a regime whereby magnetic excitations become strongly damped due to heat bath scattering. In this regime, standard perturbative techniques fail. To resolve this issue, we develop a new finite frequency, finite temperature technique for a nonlinear quantum field theory. Finally, we develop a theory of the magnetic order parameter appropriate to probe the phenomenon of *dimensional reduction*—our results provide an alternate perspective on this enigmatic problem.

In the second part of this thesis, we consider magnon Bose-condensates. To begin, we develop a consistent theoretical framework to describe the finite temperature properties of the system—our results resolve a long theoretical complication. Using the developed theory, we gain unchartered access to the magnon Bose-condensate phase diagram and subsequently uncover two new universality classes. Finally, in a related study, we predict the emergence of a long-lived, stable Higgs excitation in magnon Bose-condensates. The stability of this Higgs excitation owes to a non-trivial hybridisation mechanism.

In the final part of this thesis, we consider an aspect of deconfined quantum criticality. Inspired by numerical studies, we propose a modified quantum field theory and subsequently discover a magnetic field induced, Bose-Einstein condensate of fractionalised magnetic excitations—an apparent violation of the spin-statistics theorem. We formulate a "smoking gun" criterion to test this novel prediction.

Acknowledgements

I gratefully acknowledge Prof. Oleg Sushkov for his role as a supervisor and mentor. The enjoyable and enlightening experiences as a Ph.D. student are a result of Oleg's endless kindness and almighty command of the subject. Oleg's approach to research problems is analogous to the trajectory of a classical particle—he follows the *least-action principle*. This may seem easy, but it is an art—it requires immense clarity to avoid the allure of irrelevant information and to swerve the costly technical pitfalls in order to remain on the path of least action. Contrastingly, my approach has been analogous to a meandering quantum particle—sampling all paths from those of least action to those of unimaginable time and energy expenditure. As a Ph.D. student, it has been my task to learn techniques and digest concepts, but the real endeavor has been to adopt as much as I can of Oleg's method. I thank Oleg for his intuitive and encouraging teaching style. In the later stages of the Ph.D., he has dedicated much of his own time to help me establish my future. This goes above his duties and truly reflects his kindness. I thank Tamara and Oleg for the entertaining barbecues.

I am grateful to Sam Bladwell, Yaroslav Kharkov, Tommy Li, and Dima Miserev for the passionate physics discussions. My understanding and ideas have been shaped by this group. I especially mention Yaroslav Kharkov for our work together on research projects—any time spent working with Yaroslav has been a time of accelerated learning for me. Similarly, Sam Bladwell offered a board range of discussion topics, and I have benefited from his wisdom and good nature.

I thank Prof. Christian Ruëgg who, without any obligation whatsoever, has been very involved in my thesis pursuit—offering encouraging discussions, generating research ideas, and providing an unforgettable opportunity to present at his home institute in Switzerland. Christian has also been a huge support in helping me establish a future in physics. He is a legend.

I am grateful to Max, Finn, Lisa, and TG for being so eagerly a part of my life and for being so welcoming into theirs. I am very thankful for the times we have spent together, and for the different perspectives I've gained as a result. I thank Kendal for his wise words and unique and insightful view of the world that he has so caringly shared with me. I always enjoy discussions with Kendal and I always get a lot out of them.

I owe Hunter for a lifetime of his teachings. I enjoy Hunter's clear and resolute thinking as well as his, almost contradictory, passive nature. He is the great observer of people and their behavior, and I have done well to have him as a guide. So I am happy to acknowledge the importance of Hunter's influence on my personality and way of thinking. Moreover, more recently, more relevant... conversations with Hunter have placed our separate scientific and philosophical interests into a broader, far more abstract framework—washing away the finer details to expose a bigger picture.

I thank Tempest for helping me across all aspects of life. Temp and I share an energetic determination, but we apply ourselves in different realms. I am always benefiting from conversations with Temp—learning about where her mind/passion is taking her. And I always appreciate Temps excitement to hear what I have to say about my own direction. Temp has been the most significant and tangible support for me since the beginning of this university endeavor: Temp gave me somewhere to live when I moved to Sydney, got me my first job, got me my second job after I was fired from the first job, forcefully coordinated the furniture arrangement of my first crummy apartment, and graciously helped me move into my latest sophisticated apartment. More importantly, however, is the *intangible* support Temp offers. In many ways, Temp has been delicately shaping my life for me—always taking care to ensure I'm working toward what I want and that I'm happy. I take very seriously what Temp tells me to do. And I am very lucky to have Temp always happy to spend time with me.

This thesis is dedicated to mum and dad. They provided the tools I needed for this particular pursuit—enough discipline to be focused and clinical, but with plenty of freedom to be creative and pursue my own interests. The emotional support and stability of my childhood, as well as the great experiences and culture I have been exposed to, has helped me confidently lead the life that I want. They gave me a perspective that has kept me calm in the face of important decisions.

Dad and I have discussed many facets of research. Although the discussions are *almost* always useful, they are *certainly* always profound. So to stick to the theme, I offer a statement dad made some years ago, "I will not limit your interpretation with my own." This statement frequently resurfaces in my mind, and I have found it profoundly relevant to theoretical physics where the equations are mathematically correct, yet the physical realities they proclaim describe are open for interpretation. I am therefore reminded to tread lightly on stamping too much authority, since it may ultimately be to the detriment of the audience.

A perfectly typical conversation with mum is one that lifts me away from the narrow focus of my own research or any complication I face, provides me with a clearer perspective, and leaves me chuckling afterward. The driving force behind this lightheartedness is mum's incredible ability to understand and see far beyond the confusing and trivial issues that bother the rest of us. It is from this omnipotent vantage point that mum views the hidden correlations behind all of our experiences. I frame it in this language because my thesis explores correlated quantum matter, for which I have become an expert, yet I stand little chance of understanding correlated human behavior to the level of mum's expertise.

Finally, I thank Sophie Russell. Sophie is an endless source of encouragement, providing a transcendental enthusiasm toward this thesis—from before it started to, I imagine, long after no one has read it. Sophie's involvement has pushed me to view the thesis through the eyes of myself as a first-year physics student—with wonderment and happy disbelief. Sophie really has *energized* the entire Ph.D. experience, and I am so thankful for her willing and very essential involvement. Without Sophie, I imagine an entirely different experience, not as momentous, not as exciting.

July 2017 Harley Scammell

Contents

Chapter 1
Introduction

1.1 Preliminary Statements

In the study of quantum phases of matter and their transitions, reducing the problem to a description of the emergent phenomena has become a paradigm providing clarity in the face of otherwise unimaginable complexity. This thesis is solely concerned with *strongly-correlated electron systems*. In such quantum systems, the complexity arises due to interaction-induced correlations between the many electronic degrees of freedom—rendering exact or analytical calculations impractical or even impossible. Circumventing the complexity, one instead studies the systems emergent phenomena—emergent collective degrees of freedom, and symmetries that, upon analysis with a powerful basis set of ideas, allow characteristic aspects of the original system to be readily understood. To a good approximation, the following ideas may be regarded as that basis:

- The *order parameter* description of phase transitions and spontaneously broken symmetries. The ground state in a (spontaneously) broken-symmetry phase necessarily acquires a nonzero vacuum expectation value, the order parameter, and corresponds to a macroscopically ordered phase. This idea was introduced by Landau and Ginzburg [1].
- The *renormalisation group*, as introduced by Kenneth Wilson [2], has many implications. It provides a microscopic origin to the already successful scaling analysis popularised by Widom, Fisher, and Kadanof [3–5]. For this thesis we will understand it as a mathematical way of tracking changes in the degrees of freedom as different length/energy scales are probed. The renormalisation group together with the order parameter concept is generally known as the *Landau-Ginzburg-Wilson* paradigm of symmetry breaking and phase transitions.
- *Universality* refers to the observation that distinct physics systems display the same characteristic physical properties in the vicinity of a phase transition. The distinct physical systems may then be organised into *universality classes*. A universality

H. Scammell, *Interplay of Quantum and Statistical Fluctuations in Critical Quantum Matter*, Springer Theses,
https://doi.org/10.1007/978-3-319-97532-0_1

class is set by the symmetry properties of the order parameter and the dimensionality of the system. The universality class uniquely determines the scaling properties of a system undergoing a phase transition.

- *Quasiparticles* will be understood in this thesis to mean the low-energy fluctuations of the emergent, collective degrees of freedom of the many-body system. The emergent quasiparticles need not share the properties of the constituent particles—be it charge, spin, nor mass, yet crucially, the critical singularities observed at phase transitions are attributed to the low-energy quasiparticle degrees of freedom.

The combination of some or all of the above ideas has proven to be immensely powerful in describing the physics of metals, superfluids, classical and quantum spin systems, BCS superconductors, nuclear physics phenomenology, high-energy electroweak symmetry, etc. And has hence provided an insightful connection between many subfields of physics. Moreover, restricting ones attention to either the purely quantum ($T = 0$) or the purely classical situation, where time dynamics are irrelevant, the Landau-Ginzburg-Wilson (LGW) paradigm offers a cookbook recipe to analyse the critical observables of such systems.

Despite the enormous success of the LGW framework, scenarios arise where the standard application fails to capture the relevant physics. This thesis considers two such scenarios: First, the combined interplay of quantum and thermal fluctuations presents challenges to the above approach. Second, unconventional (exotic) phases exist with emergent fractionalised quasiparticles/excitations that appear to be forbidden by the above arguments relating to the order parameter, and instead require a modified framework. One such framework, known as *deconfined quantum criticality* [6], will be discussed in this thesis.

1.1.1 Notation and Semantics

Throughout this thesis we set $\hbar = k_B = 1$. Semantics relating to the dimensionality of the system will be understood as follows: A three-dimensional (or 3D) system refers to the spatial dimensions. A 3+1D system refers to three spatial dimensions plus a time dimension. Importantly, a 3D or three-dimensional *quantum* system will be implicitly understood to have three spatial dimensions plus a time dimension.

1.1.2 Critical Phenomena

1.1.2.1 Observables and Universality

I will speak exclusively about continuous phase transitions. Critical systems, be they quantum or classical, are identified by a diverging correlation length ξ. For a classical critical system, the phase transition is driven by thermal fluctuations and the correlation length exhibits a power law divergence as the critical temperature, T_c, is approached,

$$\xi \sim (T - T_c)^{-\nu}. \tag{1.1}$$

The exponent ν is a universal feature of the system. Other thermodynamic observables of the system, the specific-heat, susceptibility, order parameter, etc. can be expressed in terms of power law expressions in ξ. The corresponding power law exponents, the *critical exponents*, are universal features of the system belonging uniquely to the universality class. Determination of the critical exponents of classical critical systems immediately follows from the LGW technique [2].

Quantum phase transitions have garnered intense interest over the past twenty years, finding a broad range of applicability in many interesting systems. For review articles directly relevant to this thesis, we refer the reader to Refs. [7, 8]. A quantum phase transition is a critical rearrangement of the ground state driven by non-thermal fluctuations, which ultimately derive from the Heisenberg uncertainty relations. The corresponding non-thermal tuning parameter, g, may be achieved via an applied pressure, magnetic field, dopant concentration, etc. The correlation length in a quantum critical system exhibits analogous power law divergence to Eq. (1.1), but instead is controlled by g,

$$\xi \sim (g - g_c)^{-\nu}. \tag{1.2}$$

Once again, universal features of the quantum critical system are understood as power law expression in ξ. Unlike their classical counterpart, for quantum critical systems one must also introduce a temporal correlation length,

$$\xi_\tau \sim \xi^z, \tag{1.3}$$

where z is the dynamical critical exponent. The dynamical critical exponent encodes information about the dynamics of the systems quasiparticles and can be understood in terms of the *gapless* excitations at the quantum critical point (QCP). Here dispersion takes the form $\omega_{\boldsymbol{k}} \sim k^z$, and k is the momentum (as measured from the ordering wavevector). We will explicitly deal with two cases, relativistic dispersion $k = 1$, and effective non-relativistic dispersions $k = 2$. The quasiparticle excitation gap, Δ, is defined as the minimum of the excitation energy $\omega_{\boldsymbol{k}}$. Hence *gapless* excitations have a minimum excitation energy of $\min[\omega_{\boldsymbol{k}}] = 0$.

For critical quasiparticles, their excitation gap Δ is a characteristic energy scale which vanishes at the QCP as,

$$\Delta \sim \xi_\tau^{-1} \sim |g - g_c|^{\nu z}. \tag{1.4}$$

An important mapping exists between $T = 0$ quantum critical systems, and classical critical systems—a d-dimensional quantum critical system maps to an effective d+z-dimensional classical critical system. We note that (effective) dimensionality is

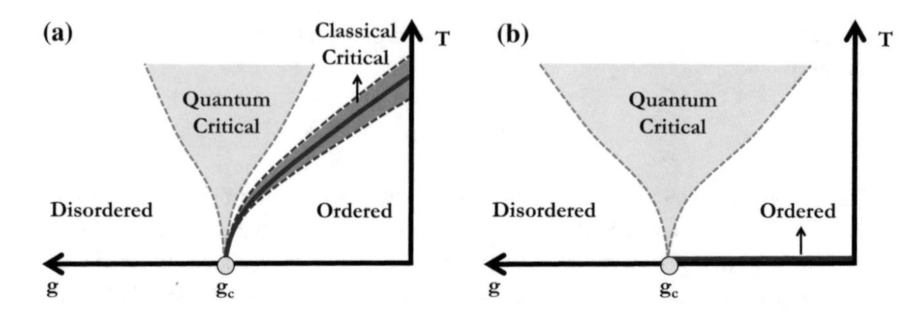

Fig. 1.1 Quantum tuning parameter g and temperature T phase diagrams. Crossover boundaries are shown by dashed lines. **a** Characteristic three-dimensional phase diagram. **b** Characteristic two-dimensional phase diagram. The ordered phase exists only at zero temperature and is indicated by the red line

a key determinant of the critical exponents, or said differently, of the scaling properties of the observables. The quantum-to-classical mapping allows for the LGW techniques [2] to be readily applied to quantum critical systems.

For both quantum and classical critical systems, there exists an upper critical dimension D_c, above which the critical exponents take on their meanfield values, and below which they become non-trivial. Precisely at the upper critical dimension exponents are meanfield, yet receive multiplicative logarithmic corrections. To be explicit, the quasiparticle gap at $d + z = D_c$ has logarithmic corrections to scaling,

$$\Delta \sim |g - g_c|^{\nu z} \ln^{\beta} |g - g_c|. \tag{1.5}$$

The exponent of the logarithm, β, is also set by the universality class. We will see in Chaps. 2 and 3, that the existence of a non-zero exponent, β, has a profound influence on the quasiparticles at the QCP.

In Fig. 1.1a and b, we present generic phase diagrams of a spontaneous symmetry breaking quantum phase transition, in three and two spatial dimensions, respectively. The broken symmetry phase is designated *ordered*, while the symmetric phase *disordered*. The QCP separating the phases is shown by the yellow point. Including temperature, the ordered phase for the three dimensional quantum system survives up to a critical transition temperature curve, solid red line, $T_c(g)$. The dashed lines, and corresponding shaded regions mark *crossover* behaviour that can be expected on the basis of general physical arguments, presented below. Crossovers, however, are not phase transitions—there are no critical singularities associated with them.

1.1.2.2 Crossovers

In the context of the quantum-to-classical mapping, introducing a nonzero temperature into a d-dimensional quantum system becomes equivalent to working with a d+z-dimensional classical system, whereby the extra z-imaginary time dimensions are truncated to have finite length $1/T$. In this sense, by adding temperature to the quantum system one must consider two length scales, $\xi_\tau = 1/\Delta$, and $L_T = 1/T$.

Tuning the relative size of the quantum ξ_τ and thermal L_T characteristic scales has novel consequences. Based on the relative size of ξ_τ / L_T, the phase diagram may be partitioned into qualitatively distinct regimes, see Fig. 1.1:

- $\xi_\tau \ll L_T$, the systems behaviour is essentially unchanged from the pure $T = 0$ case.
- $\xi_\tau \gtrsim L_T$, the system is said to be *quantum critical*. When $\xi_\tau < \infty$ the system is non-critical, however if $\xi_\tau \gtrsim L_T$ the finiteness of the quantum fluctuation length scale is masked by the length scale set by the temperature. The quantum critical behaviour, indicative of the QCP, may therefore persist into a finite volume of the phase diagram.
- $\xi_\tau \gg L_T$, the length scale set by the thermal fluctuations is drastically shorter than those of the quantum fluctuations. Viewing this as a truncation of the imaginary time axis, the vanishing ratio $L_T/\xi_\tau \to 0$ suggests a dimensional reduction. This region is referred to as *classical critical*, and universal critical indices are expected to take on the values of a classical system in d-dimensions. A detailed discussion of dimensional reduction constitutes the topic of Chap. 5.

1.1.3 Effective Quantum Field Theory

Effective quantum field theories are wonderfully practical theories capable of describing, to great accuracy, some of the most important phenomena occurring in nature. Prominent effective quantum field theories include the non-relativistic Schröedinger field theory, the four-Fermi theory of the weak nuclear force, pi-meson/chiral Lagrangian theory of the strong nuclear force, and the low energy Einstein Lagrangian as a perturbative quantum theory of gravity. The shortcoming, of course, is that effective field theories provide an accurate description of the system only at energy scales where the effective degrees of freedom and symmetries offer a reasonably faithful representation of the physics. Moreover, the effective degrees of freedom of the system are scale dependent, and hence a complete description of the physics at extended scales must be able to account for this fact.

In quantum critical systems pertaining to strongly-correlated-electrons, the effective degrees of freedom and symmetries are almost always markedly different to the microscopic degrees of freedom, i.e. the electrons. This is both an exciting and limiting feature of such systems. First, it provides the opportunity to study and observe exotic quantum fields, with unusual symmetry and dimensionality. For example, the effective degrees of freedom may display emergent Lorentz invariance, confined dimensions, global rotational symmetries, or even local gauge symmetries. Emergent, local gauge symmetries play an important role in *deconfined quantum criticality* scenario [6], to be discussed shortly. However, the effective field theory may then bare no resemblance to the underlying microscopic theory, and no hope of complete microscopic understanding is possible.

Despite any shortcomings effective quantum field theories, equipped with renormalisation group techniques [2], are capable of describing physics over sufficiently broad energy scales to remain predictive. In general, scale changes can be incorporated into the coupling constants of the field theory, whereby the coupling *constants* themselves acquire a scale dependence. We will now discuss a remarkable implication of coupling constant renormalisation, namely, *asymptotic freedom*.

1.1.4 Asymptotic Freedom

Asymptotic freedom is a generic property of 3+1 dimensional relativistic quantum field theories with a dimensionless interaction coupling constant. The argument is generic and relies on RG, the important point is that such theories sit at their upper critical dimension $D_c = 4$, and coupling constants of the action receive logarithmic scale dependent corrections. In either the ultraviolet (UV) or infrared (IR) limit, the interaction coupling vanishes logarithmically, leaving a *free* (non-interacting) quantum field theory.

Consider the interaction between gauge bosons and fermions in QED and QCD, with charges e and g, respectively. Here e is electric charge of the $U(1)$ gauge symmetry, while g is the *colour charge* of the $SU(3)$ gauge symmetry. In a perturbative, diagrammatic expansion, the combinations $\alpha_e = e^2/(4\pi)$ and $\alpha_g = g^2/(4\pi)$ are the natural expansion parameters. An RG resummation of the vertex diagrams in either theory results in the following logarithmic scale dependence of the effective interaction coupling constants [9–11],

$$\alpha_e(\Lambda) = \frac{\alpha_e(\Lambda_0)}{1 + S_e\alpha_e(\Lambda_0)\ln\left(\frac{\Lambda_0^2}{\Lambda^2}\right)}\ ; \quad \alpha_g(\Lambda) = \frac{\alpha_g(\Lambda_0)}{1 - S_g\alpha_g(\Lambda_0)\ln\left(\frac{\Lambda_0^2}{\Lambda^2}\right)}\ , \tag{1.6}$$

where Λ is the energy scale in question, say of the scattering experiment, Λ_0 is the normalisation scale typically set by experimental data at given high energy scale. The constants $S_e, S_g > 0$ are positive and are determined by combinatorics in the diagrammatic expansion.

From the explicit forms presented in (1.6), and displayed in Fig. 1.2, one sees that the coupling in QED, $\alpha_e(\Lambda)$ increases with energy scale Λ, while the QCD coupling, $\alpha_g(\Lambda)$ decreases with energy scale Λ, and ultimately vanishes in the limit $\Lambda \to \infty$. This is the asymptotic freedom [9, 10].

Asymptotic freedom was found to be a key property of an interacting theory of quarks and gluons, whereby in the limit of infinite energy the interaction coupling constant between quarks and gluons vanishes. This renders quarks to be weakly coupled at high energies, in excellent agreement with experimental data [12]. QCD subsequently became the leading theory for the structure of nucleons, with asymptotic

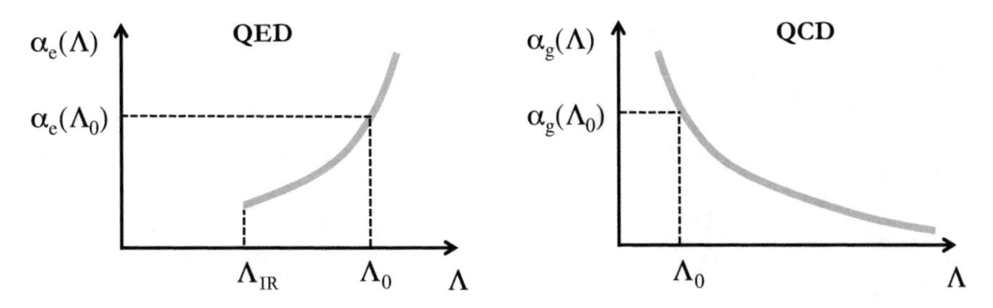

Fig. 1.2 Schematic of the scale dependent coupling constants $\alpha_e(\Lambda)$, and $\alpha_g(\Lambda)$ for QED and QCD, respectively. Normalisation scale Λ_0 is indicated. An infrared cut-off scale Λ_{IR} is present for QED due to the rest energy of the electron

freedom playing a crucial role. For the development of this theory, Gross, Wilczek, and Politzer were awarded the Nobel Prize in Physics in 2004.

In this thesis, we will be concerned with manifestations of asymptotic freedom from the perspective of thermodynamic scaling of observables, as well as in relation to dynamic scattering properties of quasiparticles.

Now that we have reviewed the essential ideas and theoretical preliminaries that form the conceptual basis of this thesis, I will now discuss the specific systems and techniques this thesis studies and applies. Within this discussion, I will hint at some research questions this thesis aims to answer—including problems with previous approaches and intuition.

1.2 Dimerised Quantum Antiferromagnets

Of primary concern to this thesis are three dimensional quantum antiferromagnets, whereby phase transitions are induced by temperature, magnetic field, or a quantum tuning parameter such as pressure. Dimerised quantum antiferromagnets are one such class that we shall refer to frequently. Dimerised quantum antiferromagnets also serve as model systems for the exotic phenomena within the *deconfined quantum criticality* scenario, to be discussed next in Sect. 1.3, and in Chap. 9.

1.2.1 Theoretical Description

Dimerised quantum antiferromagnets on bi-partite lattices are ideally suited to host quantum phase transitions. Tuning the relative strength of competing exchange couplings induces spontaneous rearrangements of the ground state. We depict one such scenario in Fig. 1.3. In the left hand side of Fig. 1.3, the spin system is dimerised into spin singlet states that form across the J' bonds. The spin singlet states have

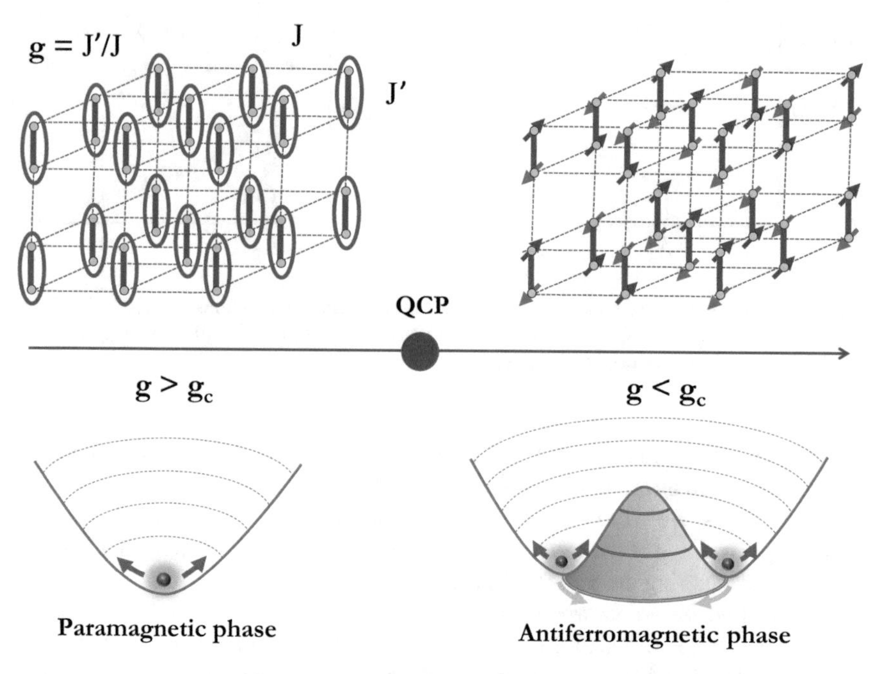

Fig. 1.3 Quantum phase transition in three dimensional dimerised quantum antiferromagnet as a function of quantum tuning parameter $g = J'/J$. (Top) Schematic quantum phase transition for a 3D dimerised lattice of $S = 1/2$ spins. Spins reside on the yellow points of the lattice. Antiferromagnetic exchange parameters J' and J on and between the dimer units, respectively, and their ratio, $g = J'/J$, controls the QPT from a quantum disordered dimer-singlet phase (left) to Néel ordered phase (right), with the QCP occurring at the critical ratio g_c. (Bottom) The transition visualised in terms of the ground state energy of the effective quantum field theory. The energy landscape is in field space; the quantum disordered/paramagnetic phase is symmetric in field space, while the ordered antiferromagnetic phase has a reduced symmetry

full $SU(2)$ spin rotational symmetry. Tuning the relative strength of the interdimer exchange coupling J, or $g = J'/J$, to some critical value g_c, the system spontaneously rearranges to an antiferromagnetic state, thereby minimising its ground state energy. The antiferromagnetically ordered state is depicted in the right hand side of Fig. 1.3. The tuning parameter g may be realised in physical systems by, for example, an external hydrostatic pressure p. In Sect. 1.2.4 we discuss a specific system whereby $g \leftrightarrow -p$.

In a low energy/low wavelength theory, the relevant, i.e. critical, degrees of freedom are captured by a field theory with three component vector $\bar{\varphi}$. Below we will obtain the low energy quantum field theory for $\bar{\varphi}$. In the symmetric phase—frequently and interchangeably referred to as the disordered or paramagnetic phase—the system has an $O(3)$ symmetry, and correspondingly, the three degrees of freedom are gapped and triply degenerate. This is depicted in Fig. 1.3 where the potential-well landscape is in field space, and the excitations/quasiparticles mass gap is represented by the curvature of the well. Spontaneous breakdown of symmetry at g_c deforms the potential-well such that two gapless (Goldstone) modes arise as oscillations in the flat directions, and one gapped Higgs mode arises as an oscillation in the curved direction.

I will now outline how to obtain the low energy quantum field theory, and then discuss extensions to the model, namely the addition of magnetic field and temperature. Note that the following derivation of the low energy quantum field theory is based upon the particular example of the simple cubic lattice of dimers, depicted in Fig. 1.3. However, in the vicinity of a quantum critical point, the microscopic details are unimportant and the same effective field theory is obtained for vastly different systems so long as their ground state symmetry and dimensionality are the same.

Consider the schematic spin-dimerised lattice depicted in Fig. 1.3. Quantum spins $S = 1/2$ reside on each yellow point, and there are two Heisenberg exchange couplings J *interdimer* shown by dotted red lines, and J' *intradimer* in solid blue lines. Denoting the spin at the upper and lower position of the bond, $\boldsymbol{S}_i^A$ and $\boldsymbol{S}_i^B$, at any given bond site i. Hence A and B refer to different sublattices. The Hamiltonian reads,

$$H = J \sum_{\langle i,j \rangle} \{\boldsymbol{S}_i^A \cdot \boldsymbol{S}_j^A + \boldsymbol{S}_i^B \cdot \boldsymbol{S}_j^B\} + J' \sum_i \boldsymbol{S}_i^A \cdot \boldsymbol{S}_i^B , \tag{1.7}$$

where first summation is over nearest neighbours, and the total number of bond sites is $N' = N/2$. The Hamiltonian (1.7) may be thought of as the fundamental theory of the quantum spin system. Of course, however, it is an approximation to the many-body Schrödinger equation. We now wish to extract a low-energy effective model from the Hamiltonian (1.7), namely, an effective quantum field theory. To derive an effective quantum field theory, we appeal to the bond-operator technique developed in Ref. [13]. For the purposes of our present goals, the bond-operator technique acts as the middle ground between the spin Hamiltonian (1.7) expressed in terms of spin operators $\boldsymbol{S}_i^A$ and $\boldsymbol{S}_i^B$, and a low-energy quantum field theory expressed in terms of the three-component quantum field $\bar{\varphi}$. The necessary connection between spin operators $\boldsymbol{S}_i^{A/B}$ and quantum field $\bar{\varphi}$ will come via the introduction of bosonic operators s_i and t_i.

To be specific, we perform the following bond-operator transformation [13],

$$\boldsymbol{S}_i^{A,B} = \frac{1}{2}(\pm s_i^\dagger t_{i,\alpha} \pm t_{i,\alpha}^\dagger s_i - i\epsilon_{\alpha,\beta,\gamma} t_{i,\beta}^\dagger t_{i,\gamma}) , \tag{1.8}$$

where the $\pm$ and the A/B refer to different sublattices, the $s_i^\dagger/s_i$ are singlet creation/annihilation operators on bond site i, and the $t_{i,\alpha}^\dagger/t_{i,\alpha}$ are triplet creation/annihilation operators on bond site i and polarisation $\alpha = \{x, y, z\}$. Importantly, the singlet and triplet states are bosonic. The bosonic quantum states created by $t_{i,\alpha}^\dagger$ are referred to as *triplons* and will be our central focus. On the other hand, the bosonic singlet states, created by $s_i^\dagger$, form a condensate and so we replace the creation/annihilation operators $s_i^\dagger/s_i$ by the condensate value $\langle s\rangle = \langle s^\dagger\rangle = \bar{s} = 1$. This constitutes a meanfield treatment, and in the last equality we set the condensate to unity for simplicity of the presentation. Now, performing the bond-operator transformation, the Hamiltonian, to quadratic order, obtains the generic form,

$$\bar{H}_2 = \sum_{\boldsymbol{k}} A_{\boldsymbol{k}} t^\dagger_{\boldsymbol{k},\alpha} t_{\boldsymbol{k},\alpha} + \frac{1}{2} B_{\boldsymbol{k}} [t^\dagger_{\boldsymbol{k},\alpha} t^\dagger_{-\boldsymbol{k},\alpha} + H.c.] \,. \tag{1.9}$$

Explicitly for the geometry of an isotropic, cubic-lattice model, one obtains,

$$A_{\boldsymbol{k}} = J' + J\bar{s}^2[\cos(\pi + k_x) + \cos(\pi + k_y) + \cos(\pi + k_z)] \,, \tag{1.10}$$

$$B_{\boldsymbol{k}} = J\bar{s}^2[\cos(\pi + k_x) + \cos(\pi + k_y) + \cos(\pi + k_z)] \,. \tag{1.11}$$

The argument of the cosines involves $(\pi, \pi, \pi) + \boldsymbol{k}$ due to the antiferromagnetic ordering at $\boldsymbol{Q}_{AFM} = (\pi, \pi, \pi)$ and the lattice parameter is set to unity $a = 1$. We are now in a position to pass to a continuum-field theory in the three component field real $\bar{\varphi}$. To do so we note that $\bar{t} = \{t_x, t_y, t_z\}$ is a three component complex field $\bar{t}^\dagger \neq \bar{t}$, and hence to account for all six degrees of freedom we must introduce two three-component field real $\bar{\varphi}$ and $\bar{\Pi}$. We make the following definition $\bar{t} = Z(\bar{\varphi} + i\bar{\Pi})$, such that $\bar{\varphi} \propto (\boldsymbol{S}_1 - \boldsymbol{S}_2)$, $\bar{\Pi} \propto (\boldsymbol{S}_1 + \boldsymbol{S}_2)$ while Z is a normalisation factor. Such an approach has been discussed in, for example, Ref. [14]. Rewriting the lattice Hamiltonian in the continuum limit, expanding to lowest order in momentum, and Fourier transforming, one obtains

$$\bar{H}_2 = Z^2 \int d^3x J\bar{s}^2 (\nabla\bar{\varphi})^2 + (J' - 6J)\bar{\varphi}^2 + J'a^2\bar{\Pi}^2 \,, \tag{1.12}$$

since a first order expansion in momenta $\boldsymbol{k}$ from $\boldsymbol{Q}_{AFM}$ gives $B_{\boldsymbol{k}} = 1/2J\bar{s}^2[\boldsymbol{k}^2 - 6]$ and $A_{\boldsymbol{k}} = J' + B_{\boldsymbol{k}}$. The corresponding Euclidean-action is found by including the Berry phase contribution, $S_B = \int d^3x d\tau \bar{t}^\dagger \partial_\tau \bar{t} = Z^2 \int d^3x d\tau 2i\bar{\Pi}\partial_\tau\bar{\varphi}$, such that,

$$S_E[\bar{\varphi}, \bar{\Pi}] = Z^2 \int d^3x d\tau \; \{2i\bar{\Pi}\partial_\tau\bar{\varphi} + J\bar{s}^2(\nabla\bar{\varphi})^2 + (J' - 6J)\bar{\varphi}^2 + J'\bar{\Pi}^2\}. \tag{1.13}$$

One immediately sees that the characteristic energy scales for the $\bar{\varphi}$ and $\bar{\Pi}$ fields are $\Delta_\varphi \sim (J' - 6J)$ and $\Delta_\Pi \sim J'$. These characteristic energy scales, Δ_φ and Δ_Π, may be understood as the energy required to excite/create either an $\bar{\varphi}$ and $\bar{\Pi}$ field. We will see shortly that we are interested in the regime where $(J' - 6J) \ll J'$, since this corresponds to the quantum critical regime. Moreover, since we are interested in a low energy effective theory, we *remove* the quantum field $\bar{\Pi}$ from the theory. Formally, *removing* $\bar{\Pi}$ involves performing a Gaussian integration of (1.13) over $\bar{\Pi}$. This provides an effective, low energy field theory in $\bar{\varphi}$, which remains faithful for energy scales $\Lambda \ll J' = \Delta_\Pi$. After straightforward manipulations, and normalisation $Z^2 = J'/2$, the low energy effective field theory is written,

$$S_E[\bar{\varphi}] = \int d\tau d^3x \frac{1}{2}\left(\partial_\tau \bar{\varphi}\right)^2 + \frac{1}{2}c^2(\nabla\bar{\varphi})^2 + \frac{1}{2}J'(J' - 6J)\bar{\varphi}^2 + S_{\text{Int}}. \quad (1.14)$$

The quadratic coupling term $J'(J' - 6J)$ drives the quantum phase transition, and it is convenient to rewrite this as $m_0^2 = \gamma^2(g - g_c)$. For this simple approximation $g_c = 6$ and $\gamma = JJ'$. Such an approximation is rather crude, mainly due to not taking into account the hard-core constraint (1.15). Ultimately, it is more convenient to leave g_c and γ as disposable, phenomenological parameters. Finally note that S_{Int} has been added to S_E in a phenomenological manner—it has not been derived from the spin Hamiltonian (1.7).

The above derivation embodies effective quantum field theories—starting from an effective action comprised of $\bar{\varphi}$ and $\bar{\Pi}$ fields and tuning to the vicinity of the QCP, $g \approx g_c$, the high energy modes $\bar{\Pi}$ are integrated out leaving just the lower energy, three component real field $\bar{\varphi}$.

The above derivation does not account for what kind of non-linear interaction term, S_{Int}, should be added. Moreover, we did not explicitly take into account the hard-core constraint,

$$s^\dagger s + t_\alpha^\dagger t_\alpha = 1\ . \quad (1.15)$$

There is some freedom to choose the interaction term, although by symmetry and stability considerations $\bar{\varphi}^4$ is a natural choice to approximate the hard-core constraint (1.15) in a low energy description. This is often called the *soft-spin* constraint. Throughout this thesis we will exclusively work with the soft-spin constraint as it naturally describes symmetry breaking phase transitions, and provides convenient access to amplitude fluctuations or *Higgs* modes.

1.2.2 Coupling to External Magnetic Field

To introduce an external constant magnetic field one performs the following transformation of the Hamiltonian (1.7),

$$H_2 \rightarrow H_2 - \sum_i \boldsymbol{B} \cdot \boldsymbol{S}_i = H_2 + iB_\alpha \sum_i \epsilon_{\alpha,\beta,\gamma} t_{i,\beta}^\dagger t_{i,\gamma}\ . \quad (1.16)$$

The corresponding adjustment to the Lagrangian density is [15–17],

$$\mathscr{L}[\bar{\varphi}, \boldsymbol{B}] = \frac{1}{2}\left(\partial_t \bar{\varphi} - \bar{\varphi} \times \boldsymbol{B}\right)^2 + \frac{1}{2}c^2(\nabla\bar{\varphi})^2 + \frac{1}{2}m_0^2\bar{\varphi}^2 + \frac{1}{4}\alpha_0^2\bar{\varphi}^4. \quad (1.17)$$

We will study the many effects induced by the presence of an external magnetic field. For now only a few comments are necessary: $\boldsymbol{B}$ breaks the full $O(3)$ rotational symmetry of the action (1.14) down to an $O(2)$ rotational symmetry about the axis

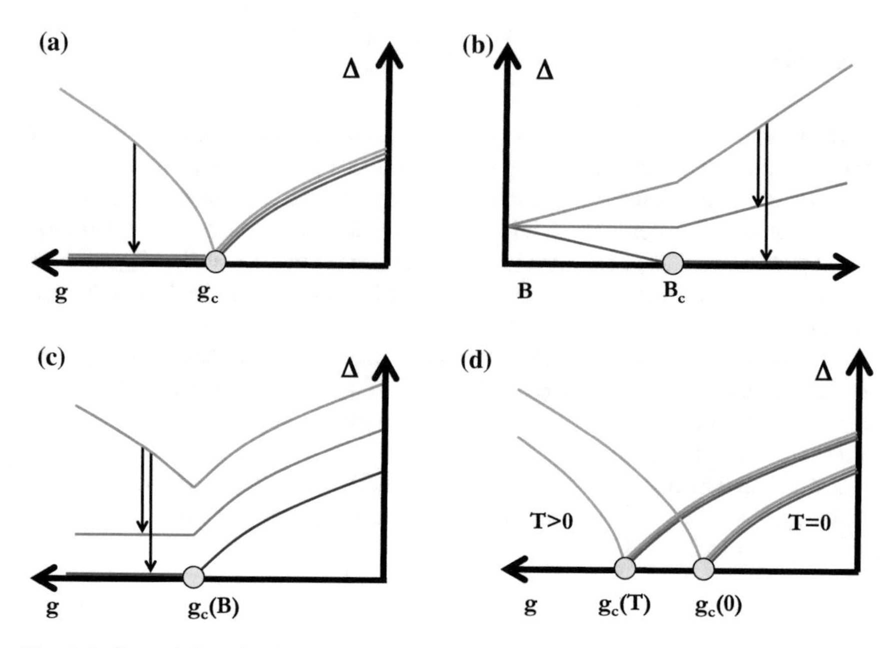

Fig. 1.4 Gaps: influence of (g, B, T). Arrows denote allowed spontaneous decay channels. Abscissa g increases from left-to-right. **a** The $T = 0$, $B = 0$ QPT driven by g. Yellow, red and green lines in the disordered phase, $g > g_c$ are degenerate triplons. In the ordered phase, $g < g_c$, yellow line represents the gapped Higgs mode, while green and red are gapless goldstone modes. The arrow indicates the possibility of spontaneous decay of Higgs modes into Goldstone modes. **b** The $T = 0$, $g = 0$ QPT driven by B. In the disordered phase, $B < B_c$, the yellow, red and green lines represent the Zeeman-split triplon modes. In the ordered phase, there is just one gapless Goldstone mode, red line, one precession mode, green line, and one Higgs mode, yellow line. Arrows indicate spontaneous decay channels of the Higgs mode. **c** The $T = 0$, $B \neq 0$ QPT driven by g. The description of the modes is identical to (**b**), except the functional dependence of the modes on g is different to that on B. **d** The $T \neq 0$, $B = 0$ QPT driven by g. The description of the modes is identical to (**a**), except now the critical point is shifted by temperature $g_c(0) \rightarrow g_c(T)$

defined by $\boldsymbol{B}$. The three degenerate modes of the symmetric phase are Zeeman split such that the excitation gaps are $\Delta_\sigma = m_0 + \sigma B$, where $\sigma = 0, \pm 1$ are the triplon polarisations. Hence only the $\sigma = -1$ mode becomes critical. Figure 1.4b depicts this situation. Upon magnetic field driven condensation, staggered magnetic order develops and lies in a plane perpendicular to the axis defined by the applied magnetic field $\boldsymbol{B}$. Within this phase, there exists one Goldstone mode, and two gapped modes. Of the gapped modes, one is an amplitude fluctuation or Higgs mode, while the other is a precession mode with rest energy set by the Larmor frequency, $g\mu_B B$, where g is the gyromagnetic factor and μ_B the Bohr magneton. Schematics of the evolution of the excitation gaps through the magnetic field driven QCP are presented in Fig. 1.4b.

The magnetic field changes the universality of the quantum phase transition—only one Goldstone mode is generated, as opposed to the $O(3)$ QPT driven by g which generates two Goldstone modes, in Fig. 1.4a.

It is worth spending a few lines commenting on the QFT (1.17). Having just one critical mode and global $O(2)$ symmetry, this effective field theory belongs to the 3+1 dimensional BEC universality class [18]. Furthermore, based on this idea, one can eliminate the higher energy modes, and extract a *lower-energy* effective theory, namely the following,

$$\mathscr{L}[\bar{\varphi}_\perp, \boldsymbol{B}] \approx B\varphi_x \partial_t \varphi_y - B\varphi_y \partial_t \varphi_x + \frac{1}{2}c^2(\nabla\bar{\varphi}_\perp)^2 + \frac{1}{2}(m_0^2 - B^2)\bar{\varphi}_\perp^2 + \frac{1}{4}\alpha_0\bar{\varphi}_\perp^4 \,. \tag{1.18}$$

Here the second order time derivatives are ignored under the assumption $B \gg \omega$, and the φ_z mode has been dropped as it is non-critical. Note $\boldsymbol{B} = B\hat{z}$. The Bose-Einstein condensation (of magnons) in dimerised quantum antiferromagnets has been considered on the basis of such an effective theory in a number of theoretical works [19–21]. Importantly, this critical theory is non-relativistic; dynamical critical exponent $z = 2$, and hence effective dimensionality is $d + z = 5$. The theory now sits above the upper critical dimension $D_c = 4$, and therefore observables do not receive logarithmic corrections. In contrast, the original field theory (1.17) receives logarithmic corrections, although there is no associated asymptotic freedom as B acts as an infrared cutoff—analogously to the role played by the electron rest energy in Fig. 1.2. Logarithmic corrections in the presence of a magnetic field will be important in Chap. 7, and I will now outline one such reason.

Under the mantra of effective QFT, many approaches to magnon Bose-condensation have relied on the *critical* Lagrangian (1.18), while certainly asymptotically correct (at vanishing energy), it has failed to describe real experimental data—the BEC critical index appeared different from theory. As such, the classification of magnetic field induced magnon condensation as belonging to the O(2) BEC universality has been an open question for $\sim$20 years [19–30]. A proposed resolution of this issue came from a paper by the present author [31], and will be the subject of Chap. 7, which argues that the combination of (i) logarithmic corrections as well as (ii) a heat bath of non-critical modes provides strong corrections to the predicted scaling near the BEC critical point. Both (i) and (ii) arise in a treatment based on Lagrangian (1.17), and are both absent in Lagrangian (1.18).

It is important now to comment on the validity of the critical theory (1.18). To this end, we provide Figs. 1.4 and 1.5 for the dependence of all three modes on tuning (g, B, T) and for the extended (g, B, T) phase diagram. Only asymptotically close the the BEC quantum critical point can one expect this modified field theory (1.18) to provide a faithful representation of the physics, namely, when $B_c \gg T$. Physically, temperature acts as a bath of low energy modes. The discarding of modes, i.e. the precession and amplitude modes, may be justified in some limited region $B \gg \omega$ at $T = 0$, however, at finite T such modes can be readily excited (thermally) and as $T \sim B$ they become relevant degrees of freedom.

The effective theory (1.18) also has no prospect of describing decay channels of non-critical modes (including the Higgs modes). Ultimately, Chaps. 7 and 8 are interested in extended regions of the phase diagram, the influence of temperature and

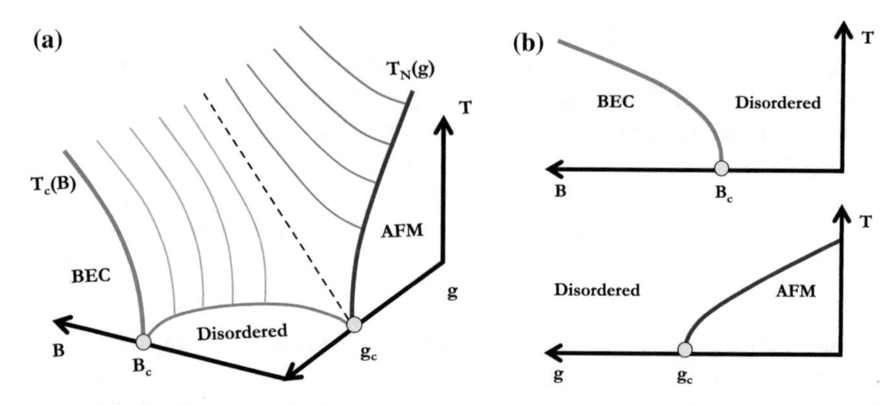

Fig. 1.5 **a** The critical surface of the (g, B, T)-phase diagram. The "disordered" phase here corresponds to the dimerised or paramagnetic phase of Fig. 1.3. **b** (Top): The critical temperature versus magnetic field in the BEC regime; $g > g_c$. (Bottom): The critical, Néel temperature versus tuning parameter g at $B = 0$. AFM order occurs for $g < g_c$

quasiparticle decay (non-critical modes), for which the full three-mode Lagrangian (1.17) is essential.

1.2.3 Research Themes and Questions

The overarching, title theme of this thesis is the *interplay of quantum and statistical fluctuations in quantum critical matter*. It is instructive to outline specific research questions commensurate with such a theme—questions that will be addressed throughout this thesis. To this end, let us give meaning to the different elements of the thesis statement. First, the quantum critical matter will be the class of dimerised quantum antiferromagnets introduced above. They will be in a regime such that only the relevant low energy degrees of freedom come from the original $O(3)$ symmetry, which may be explicitly or spontaneously broken. Second, the interplay of quantum and statistical (thermal) fluctuations will be understood and realised in many ways: (i) The interplay generates a rich phase diagram Fig. 1.5 with a critical surface and many crossover regimes. We will be interested in the role played by logarithmic scaling, and asymptotic freedom, on such a critical surface, i.e. corrections to meanfield scaling behaviour. (ii) The influence of the combined fluctuations on the energy gap of the various excitations, explicitly, the evolution of the gaps under tuning of (g, B, T). We depict this in Fig. 1.4. (iii) Quasiparticle decay—the influence of tuning handles (g, B), influence of broken vs unbroken symmetries, influence of temperature and associated heat bath scattering. Some of the possible spontaneous decay channels are indicated by black arrows in Fig. 1.4.

Within this set of problems, the influence of temperature for relative scales $T \gg \Delta$ represents a substantial challenge to theory. These scales correspond to the crossovers into the classical critical regime, and has been depicted in Fig. 1.1. Within the classical-critical regime, temperature cannot be treated as a small perturbation. Worst still, close to the classical transition (Néel temperature) a straightforward application of the QFT and RG techniques [32] generates infrared divergences in both the static observables e.g. the order parameter, and within dynamic observables e.g. the decay width of Higgs/triplon modes. Techniques going beyond perturbative QFT are called for. We develop and discuss such techniques in Chaps. 4 and 5.

1.2.4 Experimental and Numerical Realisations

As much as possible, the theoretical results of this thesis are tested against the available experimental and numerical data. We now introduce the two most prominent systems appearing in this thesis: $TlCuCl_3$, studied experimentally, and the double-cubic J–J' model, studied using quantum Monte Carlo.

Thallium copper chloride, $TlCuCl_3$, is a model material that realises an insulating, quantum magnetic system of $S = 1/2$ Cu^{2+} ions. The $S = 1/2$ spins are dimerised due to the geometry of the exchange interactions, and at ambient pressure and zero applied magnetic field, the system exhibits a quantum paramagnetic phase; spin dimers form singlet states. The structure is depicted in Fig. 1.6, where dimerisation of the $S = 1/2$ moments of the Cu^{2+} ions is indicated by the blue ovals. The system has anisotropic lattice spacings, and an easy-plane spin-orbit anisotropy, which influences excitation velocities and gaps, respectively [33–35].

Inelastic neutron scattering studies have determined that the energy gap, at ambient pressure, is $\Delta \approx 0.7$ meV [24, 36], and that the magnetic properties are three dimensional—indicated by the strong excitation dispersion in all three spatial dimensions [24, 26]. These two properties render $TlCuCl_3$ a remarkable material, allowing for experimental access to a variety of universal critical phenomena:

- Quantum $O(3)$: Remarkably, Tanaka et al. [37] found that an applied hydrostatic pressure closes the energy gap $\Delta \rightarrow 0$, and subsequently induces antiferromagnetic order. Hence hydrostatic pressure, p, plays the role of the quantum tuning parameter, g, introduced above, see e.g. Fig. 1.3. Note that $p \leftrightarrow -g$. Ignoring the influence of the spin-orbit anisotropy, the pressure induced phase transition belongs to the $D = 4$, $O(3)$ universality class. A theoretical study by the present author [35], also detailed in Chap. 2, provides strong support in favour of this universality.
- BEC: The gap, Δ, may also be closed by application of a magnetic field which, as discussed in Sect. 1.2.2, induces an antiferromagnetic order in the plane perpendicular to the applied field. This theoretical expectation has been confirmed in $TlCuCl_3$ by neutron-diffraction measurements [27]. The required critical field, $B_c \approx 5.6$ T, makes $TlCuCl_3$ one of the few known inorganic systems in which the gap may be closed by application of laboratory magnetic fields [22].

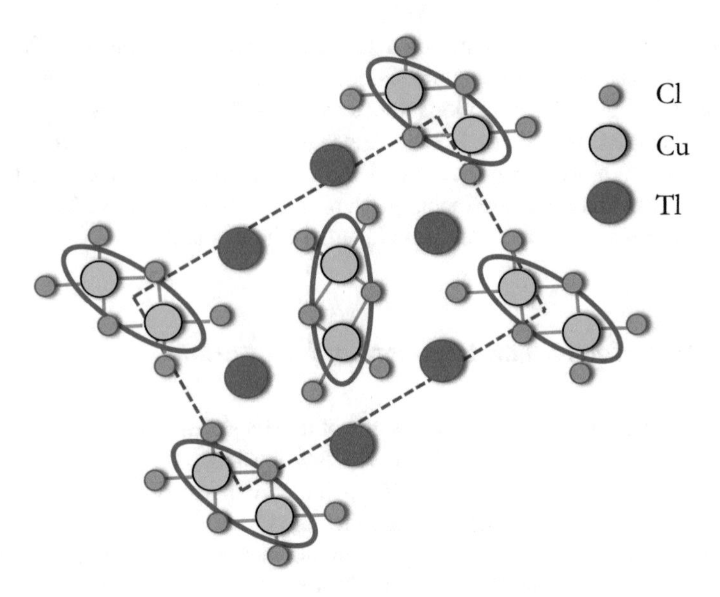

Fig. 1.6 Structure of three dimensional dimerised quantum antiferromagnet $TlCuCl_3$. Dimerisation of the $S = 1/2$ moments of the Cu^{2+} ions is indicated by the blue ovals

- Classical $O(3)$: Due to the three dimensional magnetic character of the material, the pressure induced antiferromagnetic ordered phase survives up to a non-zero Néel temperature [38]. The thermal phase transition is said to belong to the classical, or 3D, $O(3)$ universality class. This scenario has been realised by inelastic neutron scattering studies under combined pressure and temperature [39]; the studies mapped out the evolution of the mode gaps through the thermal/classical phase transition.

Hence the *smallness* of the gap, Δ, places $TlCuCl_3$ in the vicinity of the quantum critical point separating the quantum paramagnetic phase from the nearby ordered phases. And the three dimensional character allows magnetic order at non-zero temperatures. $TlCuCl_3$ therefore hosts a broad variety of critical phenomena and, accordingly, there has been a multitude of experimental [22, 26, 27, 33, 36, 39–42] and analytical [20, 21, 31, 35, 43–45] studies of this material. There also exists a group of related compounds, of which we mention $KCuCl_3$. $KCuCl_3$ hosts a similar spin dimerised structure [22–24, 40, 46–48], although, due to weaker interdimer exchange couplings [49], the excitation gap is larger $\Delta \approx 2.7$ meV, with corresponding critical magnetic field, $B_c \approx 20$ T [22, 24], being less accessible under laboratory conditions.

1.2.4.1 Numerical Systems

A convenient model system that can be analysed numerically is the double-cubic, J–J' model. This system consists of two interpenetrating cubic lattices with the

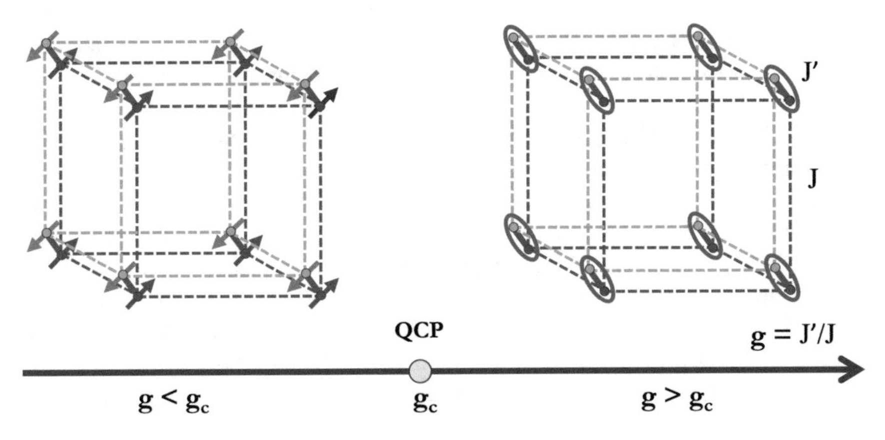

Fig. 1.7 Dimerised lattice of $S = 1/2$ spins in the $3D$ double cubic geometry. Sites of the red and light green cubic lattices are connected pairwise by dimer bonds; J' and J are antiferromagnetic Heisenberg interactions on and between the dimer units, respectively, and their ratio, $g = J'/J$, controls the QPT from a Néel ordered phase (left) to a quantum disordered dimer-singlet phase (right), with the QCP occurring at the critical ratio g_c

same antiferromagnetic interaction strength, J, connected pairwise by another antiferromagnetic interaction, J'. It serves as a representative 3D dimerised lattice with an unfrustrated geometry. The QPT occurs when the coupling ratio $g = J'/J$ is increased, changing the ground state from a Néel-ordered phase of finite staggered magnetisation to a dimer-singlet ("quantum disordered") phase, as illustrated in Fig. 1.7. The Hamiltonian takes the same form as presented in Eq. (1.7), where A and B sub-lattices now correspond to the different cubes.

Quantum Monte Carlo (QMC) simulations of this model have been performed [50–52] and have demonstrated to a high accuracy that the phase transition, driven by tuning g, belongs to the expected $d = 4$, $O(3)$ universality class. Namely, the static quantities, order parameter, Néel temperature, and dynamic quantities, triplon and Higgs gaps, have been shown to have the correct critical exponents of the multiplicative logarithmic corrections. This is a highly non-trivial issue, and will be discussed in detail in Chap. 3. It is through the latest techniques of stochastic series expansion and analytic continuation, that QMC studies [51, 52] have determined dynamic quantities.

1.3 Deconfined Quantum Criticality

We now consider an exotic scenario proposed for dimerised quantum antiferromagnets, one that requires a theoretical framework beyond a näive application of the Landau-Ginzburg-Wilson paradigm of spontaneous symmetry breaking. Explicitly, we consider the quantum phase transition between Néel antiferromagnetic states, and a *valence bond solid* (VBS) ordered state. The VBS phase is a dimerised quantum paramagnetic phase, with essentially identical features to those met in a previous Sect. 1.2. Previously the dimerised paramagnetic phase arose in the $J - J'$ model (1.7), see also Fig. 1.3, which comprised a ground state of spin singlet dimers. An essential feature that distinguishes the dimerised paramagnetic state considered above in $J - J'$ model, and the VBS state is the following: in the $J - J'$ model, the Hamiltonian explicitly breaks lattice symmetries—some exchanges are stronger than others $J' > J$. Hence, this model pins the dimers to orient in a fixed direction, say e.g. dimers form on adjacent spin sites in the $\hat{x}$-direction. On the other hand, for VBS states, the important property is that the underlying Hamiltonian *does not* break lattice symmetries, yet the formation of the VBS phase *does* break lattice symmetries. This spontaneous breaking of an underlying symmetry makes the VBS an ordered phase.

It is believed that such quantum phase transitions are in fact second-order, thus violating the expectation of a näive application of LGW theory—for the breaking of two unrelated symmetries, LGW theory generically predicts a first-order, or region of coexistence, phase transition, and the possibility of a continuous second-order transition only occurring under fine tuning of parameters.

A theoretical proposal that accommodates a continuous transition is the so called theory of *deconfined quantum criticality* (DQC) [6, 53]. In the DQC scenario the order parameters of the AFM state and the competing VBS state are not fundamental variables, instead they are comprised of fractional degrees of freedom. The fractionalised excitations, *spinons*, are spin 1/2, bosonic degrees of freedom. In the AFM phase spinons are condensed in a Higgs-like transition. Additionally, the spinons are minimally coupled to an emergent $U(1)$ gauge field. In the VBS phase the $U(1)$ gauge field confines spinon pairs to form triplons. Confinement due to gauge fields is a familiar concept from particle physics—hadronic bound states of quarks are held together by *gluon strings*, the resulting force is distance-independent, and quarks subsequently cannot be isolated [12]. At, or in the near vicinity of, the critical point separating the AFM and VBS phases, spinons undergo a confinement—deconfinement transition, whereby coupling to the gauge field becomes sufficiently weak that the spinons may be treated as individual particles. The phase diagram hosting VBS, AFM, and deconfined spinons is depicted in Fig. 1.8.

Spontaneous VBS order driven by frustration serves as a natural realisation of the VBS-AFM transition for $SU(2)$ spin systems, and is a well established theme in quantum antiferromagnetism [54]. However, the nature of the ground state and excitations in the vicinity of quantum critical points for specific models, such as the 2D square lattice frustrated Heisenberg antiferromagnet, remains somewhat controversial [55–57]. The application of unbiased numerical techniques, such as the Quantum Monte Carlo (QMC) method, in the study of frustration driven VBS order would be enlightening, however frustrated Heisenberg systems suffer the pathological *sign problem* [58], and are thus not amenable to a QMC treatment.

Instead, other models have been pursued. A four-spin exchange quantum spin model without frustration, the JQ model [59], was designed to evade the *sign problem*. The QMC method has been applied to this model, which was shown to exhibit

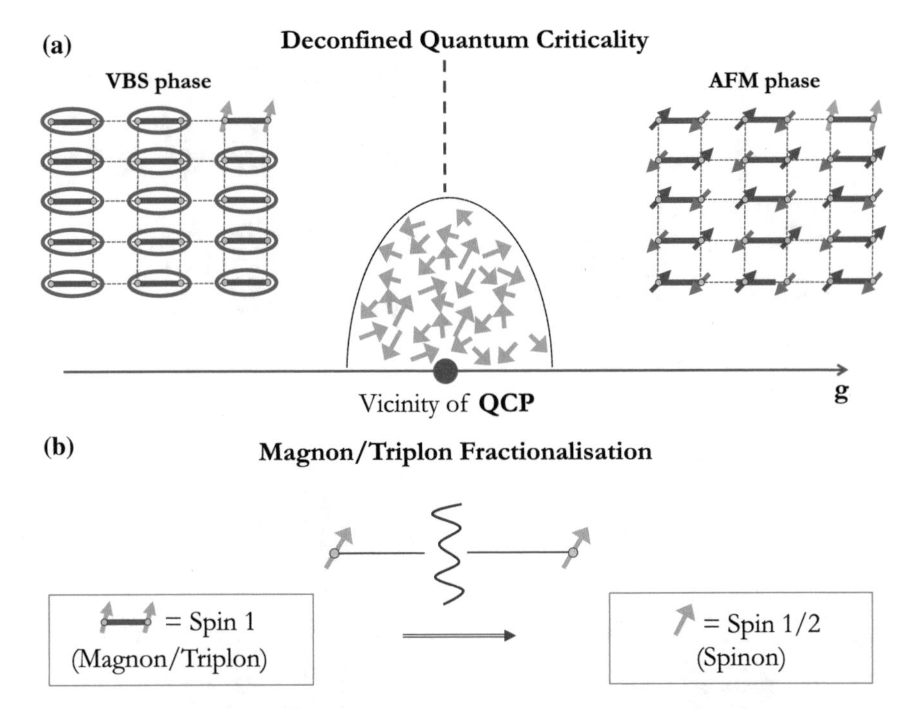

Fig. 1.8 Schematic picture of the ground state associated with **a** the AFM state, and **b** the VBS state. The arrows represent the direction of the magnetic moments. The encircled lines represent the bonds in which the spins are paired into a valence bond

columnar dimer VBS order and a magnetically ordered phase with strong hints of a deconfined QCP separating them [59]. These conclusions were later strengthened by further QMC studies [60].

1.3.1 JQ Model

For clarity and convenience, we will continue to introduce the important aspects of DQC theory, having in mind the unfrustrated, four-spin exchange model (or JQ) [59]. The JQ model is a designer Hamiltonian with the essential features characterising a VBS state, and an AFM state. It describes the competition between spin singlet projectors, with exchange energy Q, and usual antiferromagnetic Heisenberg exchange J, such that for $Q/J \gg 1$ VBS order is favoured, while for $J/Q \gg 1$ AFM order is favoured.

For spins $S = 1/2$, the model is defined using singlet projectors $P_{ij} = 1/4 - \boldsymbol{S}_i \cdot \boldsymbol{S}_j$ as,

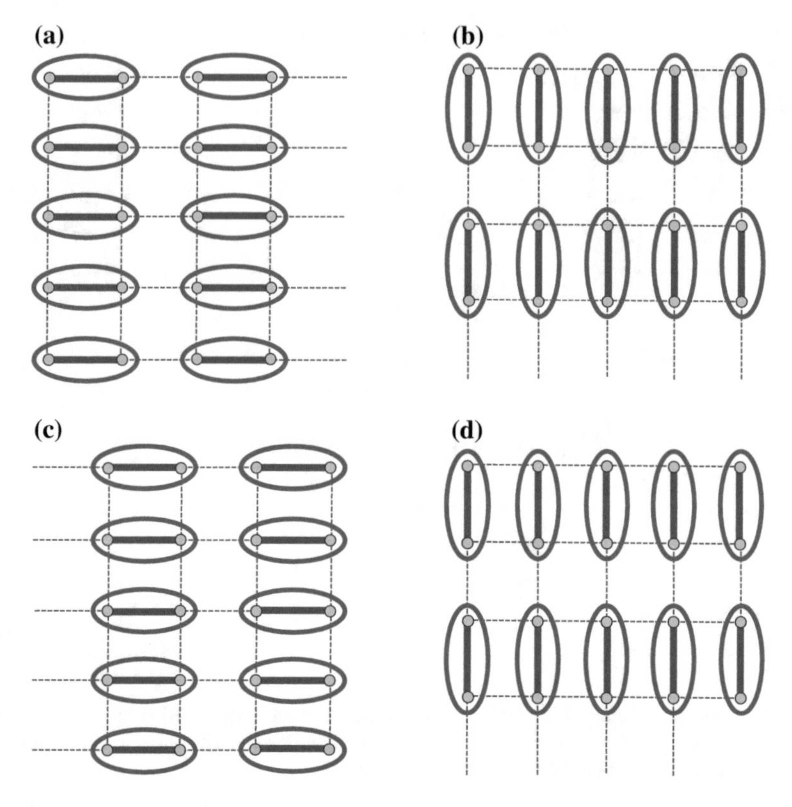

Fig. 1.9 Schematic picture of the four degenerate ground states associated with the VBS state. The ground states **a** and **b** are distinguished from each other by one horizontal lattice spacing. While, **c** and **d** are distinguished from each other by one vertical lattice spacing. The encircled lines represent the spins that are paired into a valence bond

$$H = -J \sum_{\langle ij \rangle} P_{ij} - Q \sum_{\langle ijkl \rangle} P_{ij} P_{kl} \ , \tag{1.19}$$

where $\langle ij \rangle$ denotes nearest-neighbour sites on a periodic square lattice with L^2 sites and the pairs ij and kl in $\langle ijkl \rangle$ form horizontal and vertical edges of 2×2 plaquettes. Importantly, this Hamiltonian maintains all the symmetries of the square lattice. The VBS ground state existing for $g = J/Q < g_c$, $g_c \approx 0.045$ [61], arranges in a columnar configuration as shown in e.g. Fig. 1.9a. Such a state thus breaks the translational and $\pi/2$ rotational symmetries, spontaneously. Moreover, this model hosts four degenerate ground states of dimer orientation: along $\hat{x}$, along $\hat{y}$, and a single lattice space translation of the two, as depicted in Fig. 1.9. This degeneracy means the VBS order parameter possesses discrete Z_4 symmetry. The resulting phase has no magnetic order and instead exhibits spatial ordering of the bond energy. In such a bond-ordered VBS state, the singlet projector $\langle P_{ij} \rangle$ has an expectation value that exhibits spatial structure at the VBS ordering wave-vector(s) $\boldsymbol{K}$, resultantly the

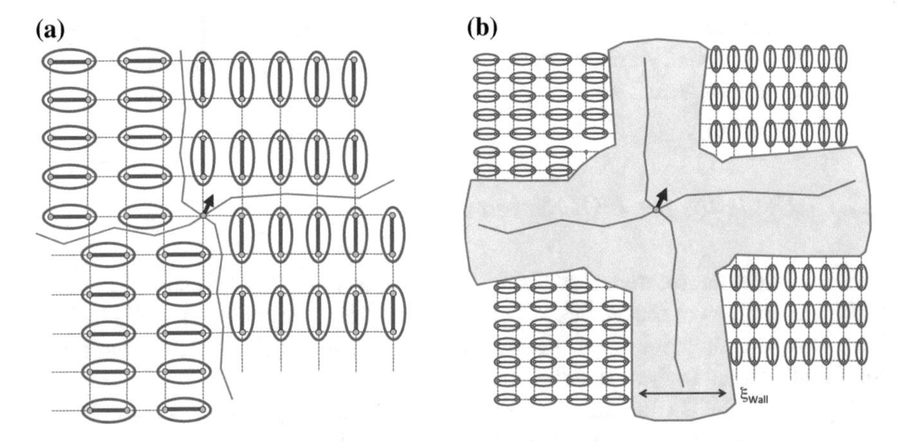

Fig. 1.10 **a** An example of elementary domains wall intersecting in the VBS state. Green lines represent the domain walls. The domain walls form a vortex in the VBS state, at the core of the vortex is an unpaired site with a free spin-1/2 moment; red arrow. **b** Thickening of the domain walls as the transition is approached. As the domain wall thickness diverges to infinity, the spinons become deconfined. The diverging length scale, ξ_{Wall}, represents the thickness of the domain wall

VBS order parameter is non-zero. The AFM state for $g > g_c$ breaks the spin-rotation symmetry identically to the AFM phase considered in the previous Sect. 1.2.

1.3.2 Spinon Schematic Model

It is instructive to provide a schematic illustration of the deconfinement of spinons. This illustration follows from the work of [62]. Consider approaching the critical point from the VBS phase. This phase has a four-fold degenerate ground state as shown in Fig. 1.9. As the phase transition is approached, patches of different, degenerate dimer orientation, and the corresponding domain walls between them, are formed at a low energy cost. Domain walls intersect at a vertex, which hosts an isolated spin $S = 1/2$ moment, a spinon (vortex), Fig. 1.10a. The presence of domain walls hinders the tunnelling events between spinons, which are ultimately responsible for confinement. The spinons have a characteristic correlation length ξ, and remain confined as long as the domain wall thickness is $\xi_{\mathrm{Wall}} < \xi$, such that tunnelling events take place between different spinons. Upon approach to the critical point, the domain wall thickens, Fig. 1.10b, and ultimately diverges $\xi_{\mathrm{Wall}} \to \infty$ at the critical point.

If the domain wall thickness diverges faster than the correlation length, $\xi_{\mathrm{Wall}} \sim \xi^{1+a}$, i.e. $a > 0$, then tunnelling events between spinons are suppressed, and they become (approximately) deconfined. According to this picture, when considering length scales $\xi < L < \xi_{\mathrm{Wall}}$, the VBS order parameter will exhibit an enhanced global symmetry, namely $Z_4 \to U(1)$. This enlarged symmetry of the VBS order parameter has been observed in QMC simulations of the JQ model [59]. Note, an emergent global $U(1)$ symmetry does not imply the DQC scenario, yet stands as important test of the present ideas.

1.3.3 Evidence for DQC Spinons

The JQ model has offered a lot of support in favour of the DQC scenario. However, QMC simulations of the model, which are necessarily performed on finite lattices, have shown scaling violations affecting many observables [61, 63–65]. It has therefore been difficult to draw definite conclusions.

Numerical QMC simulations on the JQ model strongly suggest a second order phase transition [59, 60], extensions of the model, which include for example additional (six-spin) interactions, provide additional support for a continuous QCP [66]. However it is not possible to rule out a weakly first-order transition.

The unusual scaling behaviour has been interpreted as evidence that the quantum phase transitions are in fact of first-order, as generally expected within the LGW framework [63–65]. Alternate interpretations maintain that the transitions are continuous, yet unknown mechanisms are responsible for either: strong corrections to scaling [61, 67], or fundamentally modifying scaling in an unexplained way [68, 69]. The current state of affairs, along with the puzzling aspects, are well documented in Ref. [69].

An exciting, alternate resolution to the DQC puzzle comes from a recent study [70], that performs a finite-size scaling ansatz including the two divergent length scales of the theory: the correlation length ξ of the order parameter (either in AFM or VBS, since both divergence with identical critical indices), and a second diverging length scale, ξ_{Wall}, associated with the thickness of domain walls in the VBS phase, as depicted in Fig. 1.10. The work [70] carries out simulations of the JQ model and demonstrates complete agreement with the two-length scaling hypothesis, removing all anomalous scaling corrections.

For the purpose of the work presented in Chap. 9, we are not concerned with whether the transition is of weakly first, or purely second-order in nature. We are only concerned with whether or not the spinons emerge as deconfined, fractional excitations in the vicinity of the transition, and whether they are the dominant low energy degrees of freedom—do they saturate the partition function?

Various numerical and theoretical results support existence of deconfined spinons, regardless of whether the transition is truly second-order or weakly first-order [71, 72]. The authors of Ref. [72] perform QMC simulations of the JQ model and subsequently show how the anomalies in the thermodynamical properties, specific heat, magnetic susceptibility, coherence length, can be very well accounted for by assuming spinons to be the relevant low energy degrees of freedom (as opposed to spin $S = 1$ magnons).

1.3.4 DQC Field Theory

Considering the $2+1$ dimensional DQC scenario, relating say to the square lattice model, an effective field theory has been proposed [6] to contain the relevant low energy degrees of freedom, consistent with the $U(1)$ gauge structure, and having $SU(2)$ symmetry. It has Euclidean action $S = \int d\tau d^2x \mathscr{L}$, where

$$\mathscr{L} = \{(\partial_\mu + ia_\mu)z^\dagger\}\{(\partial_\mu - ia_\mu)z\} + m^2 z^\dagger z + \frac{1}{2}\alpha(z^\dagger z)^2 + \kappa(\epsilon_{\mu\nu\gamma}\partial_\nu a_\gamma)^2 \,. \quad (1.20)$$

Here the complex, two component field z describes the spinons. They are minimally coupled to the $U(1)$ gauge field a_μ. Suppose we sit at the deconfined QCP between the AFM and VBS phases, then the transition into the AFM phase can be simply understood as the condensation of the spinon fields at $m^2 \leq 0$, giving $\langle z \rangle \neq 0$, which after the projection,

$$\bar{\varphi} = z^\dagger \bar{\sigma} z \,, \quad (1.21)$$

gives a nonzero expectation value to the real vector field $\langle \bar{\varphi} \rangle \neq 0$, and hence describes the Néel AFM ordering. Going the other direction, i.e. beginning at the deconfined critical point and approaching VBS phase, the VBS order can be understood as the proliferation of the topological defects associated with the $U(1)$ gauge field [6].

In Chap. 9, we consider a modified version of the effective field theory (1.20) and study the Bose-Einstein condensation of spinons induced by a static, applied magnetic field.

1.4 Organisation of Thesis

The rest of the thesis is organised as follows: Chap. 2 discusses non-zero temperature behaviour of observables in a three dimensional quantum critical antiferromagnet. An analysis of experimental data on $TlCuCl_3$ allows the phenomena of asymptotic freedom to be identified. Chapter 3 provides a comprehensive mapping between quantum field theory and quantum Monte Carlo simulation data. Chapter 4 analyses the non-zero temperature decay properties of triplons in three dimensional quantum antiferromagnets. A nonequilibrium technique is developed and subsequently applied to discuss the crossover region in the vicinity of the Néel temperature, where it is known that standard perturbative approaches fail. Chapter 5 discusses the interesting phenomena of *dimensional reduction* from the perspective of the order parameter. This chapter is designed to provide an alternate perspective on this interesting subject. Chapter 6 details a perturbative treatment of the non-zero temperature Bose-condensation transition capable of respecting the underlying symmetries of the action. This chapter serves as a theoretical basis for Chaps. 7 and 8. Chapter 7 predicts the emergence of two new critical indices in the extended (g, B, T)-phase diagram, as well as resolve a long standing issue with an expected critical index of the system. Chapter 8 predicts that the Higgs resonance mode in magnon-Bose condensates can be made to have vanishingly small decay width. Finally, Chap. 9 discusses the Bose-Einstein condensation of particles with half-integer spin, namely, spinons. This chapter is based on the deconfined quantum criticality framework.

References

1. Lifshitz LM, Pitaevskii LP (1995) Statistical physics: theory of the condensed state. Course of theoretical physics, vol 9. Elsevier Science & Technology
2. Wilson KG, Kogut J (1974) The renormalization group and the ϵ expansion. Phys Rep 12(2):75–199
3. Essam JW, Fisher ME (1963) Padé approximant studies of the lattice gas and Ising ferromagnet below the critical point. J Chem Phys 38(4):802–812
4. Widom B (1965) Equation of state in the neighborhood of the critical point. J Chem Phys 43(11):3898–3905
5. Kadanoff LP (1966) Scaling laws for Ising models near T_c. Physics (2):263
6. Senthil T, Vishwanath A, Balents L, Sachdev S, Fisher MPA (2004) Deconfined quantum critical points. Science 303(5663):1490–1494
7. Vojta M (2003) Quantum phase transitions. Rep Prog Phys 66(12):2069
8. Sachdev S (2008) Quantum magnetism and criticality. Nat Phys 4(3):173–185
9. Politzer HD (1973) Reliable perturbative results for strong interactions? Phys Rev Lett 30:1346–1349
10. Gross DJ, Wilczek F (1973) Ultraviolet behavior of non-abelian gauge theories. Phys Rev Lett 30:1343–1346
11. Landau LD, Abrikosov AA, Khalatnikov IM (1954) The removal of infinities in quantum electrodynamics. Dokl Akad Nauk SSSR 95:497; An asymptotic expression of the electron Green function in quantum electrodynamics 95:773; An asymptotic expression of the photon Green function in quantum electrodynamics 95:1177
12. Braun-Munzinger P, Wambach J (2009) Colloquium: phase diagram of strongly interacting matter. Rev Mod Phys 81:1031–1050
13. Sachdev S, Bhatt RN (1990) Bond-operator representation of quantum spins: mean-field theory of frustrated quantum Heisenberg antiferromagnets. Phys Rev B 41:9323–9329
14. Sachdev S (2004) Quantum phases and phase transitions of Mott insulators. Springer, Berlin, pp 381–432
15. Fisher DS (1989) Universality, low-temperature properties, and finite-size scaling in quantum antiferromagnets. Phys Rev B 39:11783–11792
16. Sachdev S (2011) Quantum phase transitions. Cambridge University Press
17. Affleck I (1991) Bose condensation in quasi-one-dimensional antiferromagnets in strong fields. Phys Rev B 43:3215–3222
18. Fisher MPA, Weichman PB, Grinstein G, Fisher DS (1989) Boson localization and the superfluid-insulator transition. Phys Rev B 40:546–570
19. Giamarchi T, Tsvelik AM (1999) Coupled ladders in a magnetic field. Phys Rev B 59:11398–11407
20. Nikuni T, Oshikawa M, Oosawa A, Tanaka H (2000) Bose-Einstein condensation of dilute magnons in $TlCuCl_3$. Phys Rev Lett 84:5868–5871
21. Kawashima N (2005) Critical properties of condensation of field-induced triplet quasiparticles. J Phys Soc Jpn 74(Suppl):145–150

22. Shiramura W, Takatsu KI, Tanaka H, Kamishima K, Takahashi M, Mitamura H, Goto T (1997) High-field magnetization processes of double spin chain systems $KCuCl_3$ and $TlCuCl_3$. J Phys Soc Jpn 66(7):1900–1903
23. Kato T, Takatsu KI, Tanaka H, Shiramura W, Mori M, Nakajima K, Kakurai K (1998) Magnetic excitations in the spin gap system $KCuCl_3$. J Phys Soc Jpn 67(3):752–754
24. Oosawa A, Takamasu T, Tatani K, Abe H, Tsujii N, Suzuki O, Tanaka H, Kido G, Kindo K (2002) Field-induced magnetic ordering in the quantum spin system $KCuCl_3$. Phys Rev B 66:104405
25. Oosawa A, Ishii M, Tanaka H (1999) Field-induced three-dimensional magnetic ordering in the spin-gap system $TlCuCl_3$. J Phys Condens Matter 11(1):265
26. Cavadini N, Heigold G, Henggeler W, Furrer A, Güdel H-U, Krämer K, Mutka H (2001) Magnetic excitations in the quantum spin system $TlCuCl_3$. Phys Rev B 63:172414
27. Tanaka H, Oosawa A, Kato T, Uekusa H, Ohashi Y, Kakurai K, Hoser A (2001) Observation of field-induced transverse nel ordering in the spin gap system $TlCuCl_3$. J Phys Soc Jpn 70(4):939–942
28. Wessel S, Olshanii M, Haas S (2001) Field-induced magnetic order in quantum spin liquids. Phys Rev Lett 87:206407
29. Nohadani O, Wessel S, Normand B, Haas S (2004) Universal scaling at field-induced magnetic phase transitions. Phys Rev B 69:220402
30. Kurita N, Tanaka H (2016) Magnetic-field- and pressure-induced quantum phase transition in $CsFeCl_3$ proved via magnetization measurements. Phys Rev B 94:104409
31. Scammell HD, Sushkov OP (2017) Multiple universalities in order-disorder magnetic phase transitions. Phys Rev B 95:094410
32. Zinn-Justin J (2002) Quantum field theory and critical phenomena. International series of monographs on physics. Clarendon Press
33. Rüegg C, Normand B, Matsumoto M, Furrer A, McMorrow DF, Krämer KW, Güdel HU, Gvasaliya SN, Mutka H, Boehm M (2008) Quantum magnets under pressure: controlling elementary excitations in $TlCuCl_3$. Phys Rev Lett 100:205701
34. Kulik Y, Sushkov OP (2011) Width of the longitudinal magnon in the vicinity of the O(3) quantum critical point. Phys Rev B 84:134418
35. Scammell HD, Sushkov OP (2015) Asymptotic freedom in quantum magnets. Phys Rev B 92:220401
36. Rüegg C, Furrer A, Sheptyakov D, Strässle T, Krämer KW, Güdel H-U, Mélési L (2004) Pressure-induced quantum phase transition in the spin-liquid $TlCuCl_3$. Phys Rev Lett 93:257201
37. Tanaka H, Goto K, Fujisawa M, Ono T, Uwatoko Y (2003) Magnetic ordering under high pressure in the quantum spin system $TlCuCl_3$. Phys B Condens Matter:329–333, (Part 2):697 – 698. Proceedings of the 23rd international conference on low temperature physics
38. Mermin ND, Wagner H (1966) Absence of ferromagnetism or antiferromagnetism in one- or two-dimensional isotropic Heisenberg models. Phys Rev Lett 17:1133–1136
39. Merchant P, Normand B, Kramer KW, Boehm M, McMorrow DF, Rüegg C (2014) Quantum and classical criticality in a dimerized quantum antiferromagnet. Nat Phys 10(5):373–379
40. Cavadini N, Rüegg C, Furrer A, Güdel H-U, Krämer K, Mutka H, Vorderwisch P (2002) Triplet excitations in low-H_c spin-gap systems $KCuCl_3$ and $TlCuCl_3$: an inelastic neutron scattering study. Phys Rev B 65:132415
41. Rüegg C, Cavadini N, Furrer A, Krämer K, Güdel HU, Vorderwisch P, Mutka H (2002) Spin dynamics in the high-field phase of quantum-critical S = 1/2 $TlCuCl_3$. Appl Phys A 74(1):s840–s842
42. Ch Ruegg N, Cavadini A, Furrer H-U, Gudel K, Kramer H, Mutka A, Wildes KH, Vorderwisch P (2003) Bose-Einstein condensation of the triplet states in the magnetic insulator $TlCuCl_3$. Nature 423(6935):62–65
43. Matsumoto M, Normand B, Rice TM, Sigrist M (2002) Magnon dispersion in the field-induced magnetically ordered phase of $TlCuCl_3$. Phys Rev Lett 89:077203

44. Matsumoto M, Normand B, Rice TM, Sigrist M (2004) Field- and pressure-induced magnetic quantum phase transitions in $TlCuCl_3$. Phys Rev B 69:054423
45. Scammell HD, Sushkov OP (2017) Nonequilibrium quantum mechanics: a "hot quantum soup" of paramagnons. Phys Rev B 95:024420
46. Takatsu KI, Shiramura W, Tanaka H (1997) Ground states of double spin chain systems $TlCuCl_3$, NH_4CuCl_3 and $KCuBr_3$. J Phys Soc Jpn 66(6):1611–1614
47. Cavadini N, Henggeler W, Furrer A, Gdel H-U, Krmer K, Mutka H (1999) Magnetic excitations in the quantum spin system $KCuCl_3$. Eur Phys J B 7(4):519–522
48. Cavadini N, Heigold G, Henggeler W, Furrer A, Gdel H-U, Krmer K, Mutka H (2000) Quantum magnetic interactions in S = 1/2 $KCuCl_3$. J Phys Condens Matter 12(25):5463
49. Saha-Dasgupta T, Valenti R (2002) Comparative study between two quantum spin systems $KCuCl_3$ and $TlCuCl_3$. Europhys Lett 60(2):309–315
50. Qin YQ, Normand B, Sandvik AW, Meng ZY (2015) Multiplicative logarithmic corrections to quantum criticality in three-dimensional dimerized antiferromagnets. Phys Rev B 92:214401
51. Qin YQ, Normand B, Sandvik AW, Meng ZY (2017) Amplitude mode in three-dimensional dimerized antiferromagnets. Phys Rev Lett 118:147207
52. Lohöfer M, Wessel S (2017) Excitation-gap scaling near quantum critical three-dimensional antiferromagnets. Phys Rev Lett 118:147206
53. Senthil T, Balents L, Sachdev S, Vishwanath A, Fisher MPA (2004) Quantum criticality beyond the Landau-Ginzburg-Wilson paradigm. Phys Rev B 70:144407
54. Sachdev S, Read N (1991) Large n expansion for frustrated and doped quantum antiferromagnets. Int J Mod Phys B 05(01n02):219–249
55. Singh RRP, Weihong Z, Hamer CJ, Oitmaa J (1999) Dimer order with striped correlations in the J_1-J_2 Heisenberg model. Phys Rev B 60:7278–7283
56. Capriotti L, Becca F, Parola A, Sorella S (2001) Resonating valence bond wave functions for strongly frustrated spin systems. Phys Rev Lett 87:097201
57. Mambrini M, Läuchli A, Poilblanc D, Mila F (2006) Plaquette valence-bond crystal in the frustrated Heisenberg quantum antiferromagnet on the square lattice. Phys Rev B 74:144422
58. Henelius P, Sandvik AW (2000) Sign problem in Monte Carlo simulations of frustrated quantum spin systems. Phys Rev B 62:1102–1113
59. Sandvik AW (2007) Evidence for deconfined quantum criticality in a two-dimensional Heisenberg model with four-spin interactions. Phys Rev Lett 98:227202
60. Melko RG, Kaul RK (2008) Scaling in the fan of an unconventional quantum critical point. Phys Rev Lett 100:017203
61. Sandvik AW (2010) Continuous quantum phase transition between an antiferromagnet and a valence-bond solid in two dimensions: evidence for logarithmic corrections to scaling. Phys Rev Lett 104:177201
62. Levin M, Senthil T (2004) Deconfined quantum criticality and Néel order via dimer disorder. Phys Rev B 70:220403
63. Jiang F-J, Nyfeler M, Chandrasekharan S, Wiese U-J (2008) From an antiferromagnet to a valence bond solid: evidence for a first-order phase transition. J Stat Mech Theory Exp 2008(02):P02009
64. Kuklov AB, Matsumoto M, Prokof'ev NV, Svistunov BV, Troyer M (2008) Deconfined criticality: generic first-order transition in the SU(2) symmetry case. Phys Rev Lett 101:050405
65. Chen K, Huang Y, Deng Y, Kuklov AB, Prokof'ev NV, Svistunov BV (2013) Deconfined criticality flow in the Heisenberg model with ring-exchange interactions. Phys Rev Lett 110:185701
66. Lou J, Sandvik AW, Kawashima N (2009) Antiferromagnetic to valence-bond-solid transitions in two-dimensional SU(n) Heisenberg models with multispin interactions. Phys Rev B 80:180414
67. Bartosch L (2013) Corrections to scaling in the critical theory of deconfined criticality. Phys Rev B 88:195140
68. Kaul RK (2011) Quantum criticality in SU(3) and SU(4) antiferromagnets. Phys Rev B 84:054407

69. Nahum A, Chalker JT, Serna P, Ortuño M, Somoza AM (2015) Deconfined quantum criticality, scaling violations, and classical loop models. Phys Rev X 5:041048
70. Shao H, Guo W, Sandvik AW (2016) Quantum criticality with two length scales. Science 352(6282):213–216
71. Kotov VN, Yao D-X, Neto AHC, Campbell DK (2009) Quantum phase transition in the four-spin exchange antiferromagnet. Phys Rev B 80:174403
72. Sandvik AW, Kotov VN, Sushkov OP (2011) Thermodynamics of a gas of deconfined bosonic spinons in two dimensions. Phys Rev Lett 106:207203

Chapter 2
Asymptotic Freedom in Quantum Magnets

Abstract In a quantum field theoretic description of three dimensional quantum antiferromagnets, the magnetic quantum critical point is expected to exhibit free quasiparticles due to the vanishing of the interaction coupling constant—an effect known as asymptotic freedom. Despite this expectation, the asymptotic vanishing of the coupling constant has never been observed such systems. In this chapter, we establish the existence, as well as explore the implications, of asymptotic freedom in the setting of three dimensional quantum antiferromagnets.

2.1 Introduction

Relativistic quantum field theories, at the upper critical dimension (3D + time), share an important, common feature—logarithmic scale dependence of the interaction coupling constant. In Quantum Chromodynamics, the interaction coupling constant logarithmically decays at high energies (short distances). Ultimately, at infinite energies particles do not interact—this is ultraviolet asymptotic freedom [1, 2]. In this case, the ultraviolet asymptotic freedom is due to non-abelian gauge fields, which act as an anti-screening mechanism. In the case of non-gauge quantum field theories or in abelian gauge theories, the coupling constant decays logarithmically in the low energy limit [3]. To distinguish between the ultraviolet case known to Quantum Chromodynamics, we will call this phenomena "infrared asymptotic freedom". However, usually this decay is terminated due to a natural low energy cutoff in the system. For example, in Quantum Electrodynamics the rest energy of the electron acts as the natural cutoff, preventing the occurrence of infrared asymptotic freedom. On the other hand, if the relativistic quantum field theory is tuned to a quantum critical point, such that all energy scales vanish, then asymptotic freedom is expected. The present work follows this line of reasoning: we search for the fingerprints of infrared asymptotic freedom within 3D quantum antiferromagnets in the vicinity of a QCP.

H. Scammell, *Interplay of Quantum and Statistical Fluctuations in Critical Quantum Matter*, Springer Theses,
https://doi.org/10.1007/978-3-319-97532-0_2

We explicitly consider the 3D quantum antiferromagnet $TlCuCl_3$, which can be driven through a QCP—separating a magnetically disordered from a magnetically ordered phase—by the application of an external hydrostatic pressure [4]. This provides a unique opportunity to study the physics described above. The low energy logarithmic behaviour at a QCP can, in principle, be pinned down even at zero temperature [5]. However, the existing zero-temperature experimental data are insufficient to pin down the logarithmic scaling. Instead combined zero and nonzero temperature data on $TlCuCl_3$ [6–8] provide additional information on the scaling behaviour of observables near the QCP, and hence provide an excellent opportunity to search for fingerprints of asymptotic freedom. In order to perform this search, we first develop a theory of the QCP which accounts for both quantum and thermal fluctuations. Having developed the appropriate, we can reliably compare the theoretical predictions with experimental data.

Before turning to the theoretical details, we outline the phase diagram and relevant observables of the system. The phase diagram of the dimerised 3D quantum antiferromagnet $TlCuCl_3$ is shown in the vertical panel of Fig. 2.1a. The disordered quantum phase consists of an array of spin dimers (a spin dimer consists of two spins arranged in the spin singlet state), and the ordered quantum state has a long range Néel order as illustrated in Fig. 2.1a. The Néel temperature curve (red line) separates ordered and disordered phases, with the QCP indicated by the yellow dot.

Excitations in the disordered phase, triplons, are gapped. These are triplet excitations of spin dimers, Fig. 2.1b(i). Note, for clarity we use terminology "triplon" instead of magnon/paramagnon, even though triplon is typically reserved for magnons arising in the context bond-operator theory. There are two kinds of excitations in the ordered phase; gapped longitudinal Higgs and gapless Goldstone excitations. They are illustrated in Fig. 2.1b(ii) and (iii). The horizontal panel in Fig. 2.1a displays excitation gaps versus pressure at zero temperature.

Overall the experimental data [6–8] provide the following information: (i) Néel temperature versus pressure, (ii) magnetic excitation gap in the disordered phase for various temperatures and pressures, (iii) Higgs magnon excitation gap in the antiferromagnetic phase for various temperatures and pressures, (iv) magnetic excitation width (lifetime) for various temperatures and pressures. To establish the existence of asymptotic freedom, we do not use the width data. However, we fully use the data from points (i), (ii) and (iii). Having established asymptotic freedom, we perform a detailed analysis of its influence on the decay width.

Finally, we mention that there is a small spin-orbit anisotropy in $TlCuCl_3$ which gaps one of the "Goldstone" modes in the antiferromagnetic phase. This implies that the number of dynamic degrees of freedom change from 3 at high energy to 2 at very low energy. We neglect this effect throughout the main text, but in Appendix A we explicitly show that the presence of the anisotropy does not influence the major conclusions.

The remainder of this chapter is organised as follows: Sect. 2.2 introduces the quantum field theoretic description of the system and the corresponding key observables. Section 2.3 directly compares the observables from quantum field theory and experimental data on $TlCuCl_3$ to demonstrate the existence of asymptotic freedom.

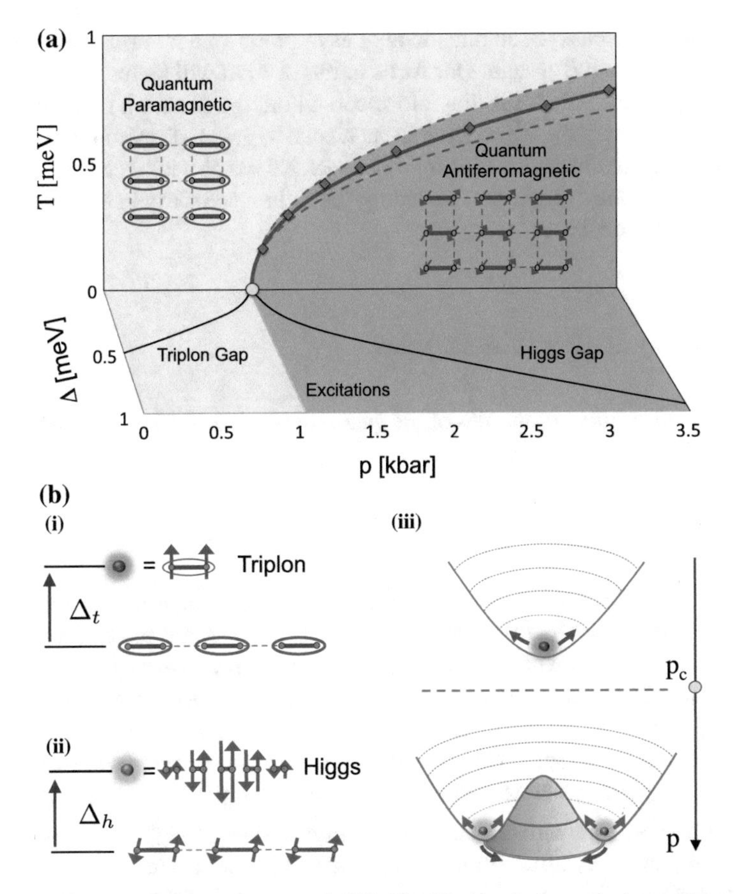

Fig. 2.1 **a** The phase/energy diagram of $TlCuCl_3$. The vertical panel shows the pressure-temperature phase diagram, the Néel temperature curve separates magnetically ordered and magnetically disordered phases. The light red band around the Néel curve indicates the region of dimensional crossover. Points show experimental data from Ref. [7]. The horizontal panel shows both the triplon gap Δ_t in the paramagnetic phase and the Higgs magnon gap Δ_H in the antiferromagnetic phase versus pressure at zero temperature. **b** Excitations of a dimerised quantum antiferromagnet. Panel (i) illustrates the triple degenerate gapped triplon excitations. Panel (ii) is a cartoon schematic of the gapped longitudinal (Higgs) excitation. Panel (iii) illustrates the quantum phase transition; the strength of the interactions in either phase is depicted by the steepness of the well. Within the ordered phase the 'Mexican hat' potential has a flat direction which supports the gapless Goldstone excitations (red arrows). Precisely at the QCP (dashed line), all directions flatten—the Higgs and triplon excitations become gapless and non-interacting i.e. asymptotically free

In Sect. 2.4 we discuss the implications of asymptotic freedom for other properties and observables in the system. Our focus in this section will be on properties of the mass gap, and decay width taking into account both the quantum running coupling and statistical heat bath. We will use these results to gain a deeper understanding of the region of quantum criticality. Finally in Sect. 2.5 we provide a comparison of our results for the decay properties, including the influence of asymptotic freedom, to experiment data on $TlCuCl_3$.

2.2 Theory I: Key Observables

To describe the critical properties of the quantum phase transition, we work with the following Lagrangian [9, 10],

$$\mathscr{L} = \frac{1}{2}\partial_\mu\vec{\varphi}\partial^\mu\vec{\varphi} - \frac{1}{2}m_0^2\vec{\varphi}^{\,2} - \frac{1}{4}\alpha_0[\vec{\varphi}^{\,2}]^2. \tag{2.1}$$

The three component vector field $\vec{\varphi}$ describes the critical excitations, and the derivatives are $\partial_\mu = (\partial_0, c\bar{\nabla})$ and we set $c = 1$. We assume the linear expansion $m_0^2(p) = \gamma^2(p_c - p)$, where $\gamma^2 > 0$ is a coefficient, p is the applied pressure and $p = p_c$ is the QCP. Hence the varying the pressure relative to p_c leads to two distinct phases; (i) at $p < p_c$, $m_0^2 > 0$, and the classical expectation value of the field is zero $\varphi_c^2 = 0$. This is the magnetically disordered phase (symmetric phase), which hosts gapped and triply degenerate excitations—the triplons. (ii) At $p > p_c$, $m_0^2 < 0$, and the field obtains a non-zero classical expectation value $\varphi_c^2 = \frac{|m_0^2|}{\alpha_0}$. This describes the magnetically ordered, antiferromagnetic phase (broken symmetry phase). Due to the spontaneous symmetry breaking, two gapless transverse (Goldstone [11]) excitations emerge, as well as one gapped longitudinal (Higgs) excitation. At the meanfield level, one finds that the Higgs gap/triplon gap$= \sqrt{2}$ [12], explicitly $\Delta_t(p) = m_0(p)$ and $\Delta_h(p) = \sqrt{2}|m_0(p)|$.

2.2.1 *Quantum and Statistical Fluctuations*

We now wish to take into account quantum and statistical (thermal) fluctuations. We consider the one-loop fluctuations corrections, which are represented diagrammatically in Fig. 2.2; they are the vertex and self-energy. By application of the renormalization group (RG), it is well known that the vertex corrections result in a logarithmic scale dependence of the coupling constant α_Λ, see e.g. Ref. [13] or Appendix A,

$$\alpha_\Lambda = \frac{\alpha_0}{1 + \frac{(N+8)\alpha_0}{8\pi^2}\ln\left(\Lambda_0/\Lambda\right)}. \tag{2.2}$$

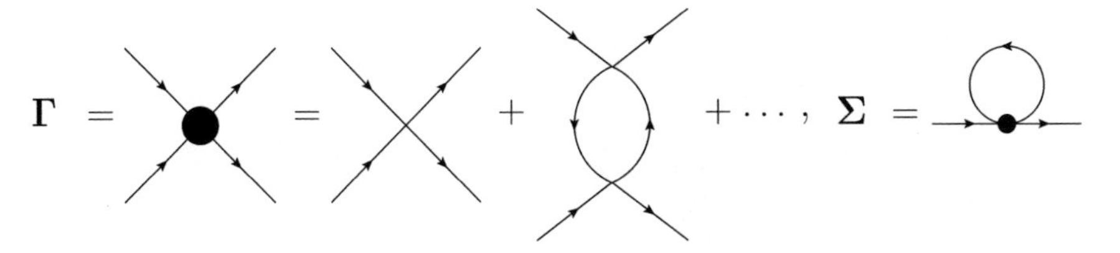

Fig. 2.2 Diagrams for the vertex Γ and self-energy Σ

We will henceforth refer to this as the *running coupling constant*. Here Λ is the energy/momentum scale, Λ_0 is the normalisation point, $\alpha_{\Lambda_0} = \alpha_0$, and $N = 3$ corresponds to the three components of the vector $\vec{\varphi}$. Since Eq. (2.2) has been obtained within the one-loop RG, we must have $\alpha_0/8\pi \ll 1$ for a reliable perturbative expansion. We see that the running coupling has a pole at $\Lambda = \Lambda_L = \Lambda_0 e^{8\pi^2/(N+8)\alpha_0}$. This is the famous Landau pole much debated in quantum field theory [3]. Of course one cannot rely one-loop RG results to adequately discuss Landau pole effects, however, we will see later that quantum magnets can shed light on the problem.

In Fig. 2.3a, we plot the running coupling constant as a function of pressure at zero temperature. As explained, the coupling constant vanishes at the QCP—this is the infrared asymptotic freedom. Figure 2.3a represents one of our central results. In order to plot Fig. 2.3a, we have input explicit parameters $\{p_c, \gamma, \alpha_0\}$, below we explain how they were obtained.

The scale dependence of the mass and of the staggered magnetisation (order parameter) are obtained by considering both the vertex and self-energy diagrams of Fig. 2.2. At zero temperature, they are well known [13], (for an explicit derivation see Appendix A)

$$m^2(p, \Lambda) = \gamma^2(p_c - p)\left[\frac{\alpha_\Lambda}{\alpha_0}\right]^{\frac{N+2}{N+8}} \tag{2.3}$$

$$\varphi_c^2(p, \Lambda) = \frac{\gamma^2(p - p_c)}{\alpha_0}\left[\frac{\alpha_0}{\alpha_\Lambda}\right]^{\frac{6}{N+8}}. \tag{2.4}$$

At zero temperature the running energy scale is set equal to the mass gap $\Lambda = m$.

We now extend the above results to nonzero temperature, i.e. taking into account statistical fluctuations, and obtain the following expressions (details presented in Appendix A)

$$\Delta_t^2(p, T, \Lambda) = \gamma^2(p_c - p)\left[\frac{\alpha_\Lambda}{\alpha_0}\right]^{\frac{N+2}{N+8}} + (N+2)\alpha_\Lambda \sum_{\mathbf{k}} \frac{1/\omega_{\boldsymbol{k}}}{e^{\frac{\omega_{\boldsymbol{k}}}{T}} - 1}. \tag{2.5}$$

Now the running energy scale is set by $\Lambda = \max\{\Delta_t, T\}$. However, hidden in the above formulation is the triplon dispersion $\omega_{\boldsymbol{k}}$. We claim that the naive on-mass-shell

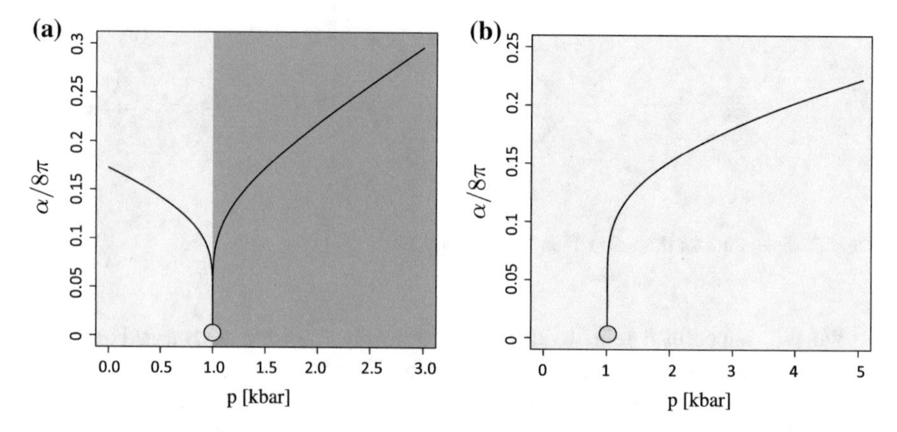

Fig. 2.3 **a** Zero temperature running coupling constant versus pressure in $TlCuCl_3$. The constant vanishes at the QCP (yellow point). **b** Running coupling constant versus pressure along the Néel temperature curve. Unlike panel **a** where the temperature is zero, in this case $T = T_N(p)$. According to Eqs. (2.5) and (2.7), the infrared cutoff in Eq. (2.2) for α_Λ is $\Lambda = T_N(p)$. The QCP is again marked by the yellow dot

dispersion $\omega_{\boldsymbol{k}} = \sqrt{\boldsymbol{k}^2 + \Delta_t^2}$ is incomplete because at small $\boldsymbol{k}$ and close to the Néel temperature where $\Delta_t \to 0$ the linewidth Γ_t (temperature broadening) becomes larger than the gap. Physically the inequality $\Gamma_t > \Delta_t$ is an indication of the dimensional crossover, $4D \to 3D$. Sufficiently close to the Néel temperature, critical indices take the 3D classical values. A detailed analysis of the triplon decay linewidth Γ_t is presented in Chap. 4, while in Chaps. 4 and 5 we discuss the $4D \to 3D$ dimensional crossover problem. For the present chapter, to account for the decay linewidth and dimensional crossover we take the following ansatz $\omega_{\boldsymbol{k}} = \sqrt{\boldsymbol{k}^2 + \Delta_t^2 + \Gamma_t^2}$. We also note that this damping-renormalized dispersion is consistent with that used for the analysis of spectral data for spin excitations in $TlCuCl_3$ by Ref. [6]. Of course the modified dispersion is not sufficient to fully describe the dimensional crossover, but it is sufficient for the purposes of the present work. The line broadening we take directly from experiment, $\Gamma_t = \xi T$, where $\xi \approx 0.15$ [6]. Later, in Sect. 2.4.2, we obtain analytic expressions (2.19) and (2.20) which justify the linear-in-T linewidth $\Gamma_t = \xi T$. Later, in Sect. 2.4.2 we will explicitly motivate this ansatz.

To find the Néel temperature as function of pressure, $T_N(p)$, we solve Eq. (2.5) with $\Delta_t(p, T_N) = 0$, which gives

$$T_N(p)^2 = \frac{\gamma^2(p - p_c)}{(N+2)\alpha_0 \sum_{\mathbf{y}} \frac{1/\omega_y}{(e^{\omega_y} - 1)}} \left[\frac{\alpha_0}{\alpha_\Lambda}\right]^{\frac{6}{N+8}}, \tag{2.6}$$

where $\omega_y = \sqrt{\boldsymbol{y}^2 + (\Gamma/T_N)^2} = \sqrt{\boldsymbol{y}^2 + \xi^2}$. We note that the critical exponent of the magnetisation in Eq. (2.4) and the Néel temperature (2.6) are identical, which agrees with the recent Quantum Monte Carlo simulations [14].

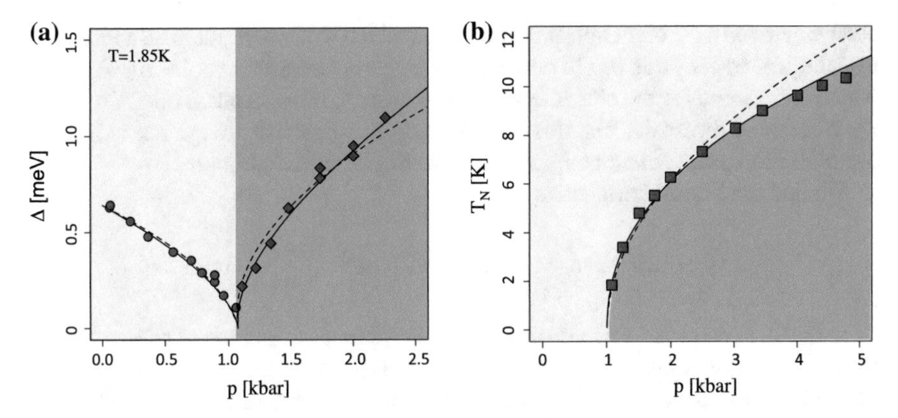

Fig. 2.4 **a** Triplon and Higgs gaps versus presssure at temperature T = 1.85 K. Points show experimental data from Ref. [8]. **b** Néel temperature versus pressure. Points show experimental data from Ref. [7]. In both plots the solid and dashed curves are quantum field theory fits with and without account of the running coupling constant, respectively

The temperature dependence of the Higgs gap in the ordered phase is found to be,

$$\begin{aligned}\Delta_H^2(p, T, \Lambda) &= 2\left\{\gamma^2(p-p_c)\left[\frac{\alpha_\Lambda}{\alpha_0}\right]^{\frac{N+2}{N+8}}\right. \\ &\left. -(N-1)\alpha_\Lambda\sum_{\mathbf{k}}\frac{1/k}{e^{\frac{k}{T}}-1} - 3\alpha_\Lambda\sum_{\mathbf{k}}\frac{1/\omega_{\boldsymbol{k}}}{e^{\frac{\omega_k}{T}}-1}\right\}.\end{aligned} \tag{2.7}$$

Again, $\Lambda = \max\{\Delta_H, T\}$, $\omega_{\boldsymbol{k}} = \sqrt{\boldsymbol{k}^2 + \Delta_H^2 + \Gamma_H^2}$, and $\Gamma_H = \zeta T$. We take $\zeta \approx 0.3$, which guarrantees that the Néel temperature determined from$\Delta_t = 0$, Eq. (2.5), is identical to that determined from $\Delta_H = 0$, Eq. (2.7). The broadening coefficient $\zeta \approx 0.3$ is consistent with data on $TlCuCl_3$ [6]. The Goldstone modes remain massless even in the presence of quantum and statistical fluctuations; being massless is a symmetry requirement.

2.3 Demonstration of Asymptotic Freedom

Having established the one-loop renormalized mass gaps and Néel temperature, we are now ready to compare with experimental data on $TlCuCl_3$ and to demonstrate asymptotic freedom. We have three fitting parameters at our disposal, the critical pressure p_c, the coefficient γ, and the coupling constant α_0. Note that $\alpha_0 = \alpha_\Lambda(\Lambda_0)$ is the running coupling constant evaluated at the normalisation point Λ_0. This means that the normalisation point is itself not an extra fitting parameter, since one can always demand a different normalisation point which will just result in an appropriate rescaling of the coupling α_0. To proceed, we set the normalisation point $\Lambda_0 = 1$ meV, which corresponds to the characteristic upper scale in the data taken on $TlCuCl_3$ [6–8]. We remind the reader that this choice is arbitrary, one can always use a different normalisation point with an appropriate rescaling of the coupling α_0.

We find the best fit parameters

$$p_c = 1.01 \text{ kbar}, \ \gamma = 0.68 \text{ meV/kbar}^{1/2}, \ \frac{\alpha_0}{8\pi} = 0.23\,. \tag{2.8}$$

In Fig. 2.4a we plot the triplon and Higgs mass gaps as a function of pressure and at fixed $T = 1.85$ K. Figure 2.4b shows the Néel temperature as a function of pressure. The points show data taken from Refs. [6–8]. Our primary goal is to demonstrate the existence of the running coupling constant within the physical system $TlCuCl_3$. To this end, we plot our theoretical results, i.e. Eqs. (2.5), (2.6) and (2.7) with parameters (2.8), both including and excluding the logarithmic running of the coupling constant; solid curves include running, while dashed exclude the running by setting $\alpha_\Lambda = \alpha_0$. We deduce that the running coupling constant plays an important role in describing the data, and hence conclude that this intriguing mathematical feature is demonstrable within $TlCuCl_3$.

Finally, we can naively extract the position of the Landau pole for the obtained parameters (2.8), $\Lambda_L = \Lambda_0 e^{8\pi^2/11\alpha_0} \approx 3.5$meV. This energy is higher than the experimentally studied regime and is comparable with expected ultraviolet cutoff, related to the lattice spacing. Future experimental studies in this energy range maybe be able to explore some aspects relations to the Landau pole, i.e. strong coupling phenomena. Alternatively, similar phenomena may be addressed in Quantum Monte Carlo studies of appropriate dimerised spin-lattice models.

2.4 Further Implications of Asymptotic Freedom

In this section we discuss the implications of asymptotic freedom for other properties and observables in the system. Our focus in this section will be on properties of the mass gap, and decay width taking into account both the quantum running coupling and the statistical heat bath. We will use these results to gain a deeper understanding of the various regimes of the disordered phase—in particular the region of quantum criticality. The remaining results relate solely to the disordered side of the phase diagram. Henceforth we will refer to the quasiparticles in the disordered phase as *paramagnons*.

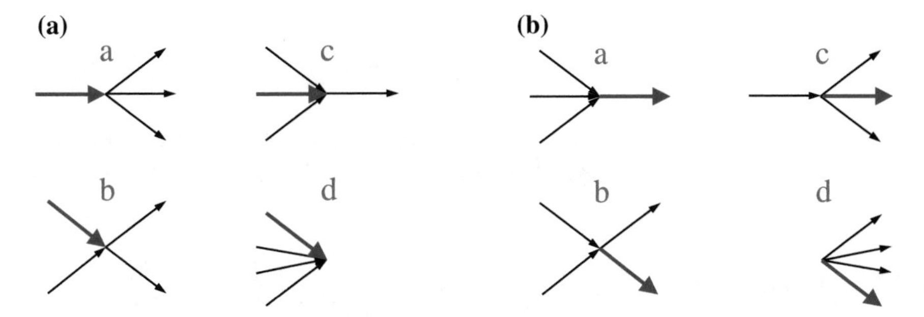

Fig. 2.5 **a** Decay diagrams for a paramagnon. The thick blue line represents the probe paramagnon and thin black lines represent the heat bath paramagnons. **b** Diagrams corresponding to pumping (inverse processes) to the paramagnon state. The thick blue line represents the probe paramagnon and thin black lines represent the heat bath paramagnons

2.4.1 Scattering from the Heat Bath

Let us consider the possible decay (or scattering) processes available to a paramagnon. Figure 2.5 provides a schematic of the decay channels available to a given probe paramagnon, represented by the blue arrows in Fig. 2.5. All such channels follow from the contact interaction term, $\alpha\vec{\varphi}^4$, in the Lagrangian (2.1). Figure 2.5a(a) shows the probe magnon spontaneously decaying into three paramagnons. Due to the heat bath, the probe paramagnon can also scatter of a heat bath paramagnon, which is often called a *Raman process*, and is shown in Fig. 2.5a(b). Moreover, there are fusion processes with two or even three heat bath paramagnons, Figs. 2.5a(c) and (d).

It is worth noting that processes Fig. 2.5a(a, c, and d) are kinematically forbidden for on-mass-shell paramagnons [15]. However, once significant deviations from the on-mass-shell condition are observed, such processes have a nonzero contribution and must be included in the analysis. This occurs, for instance, in the vicinity of the Néel temperature and will be the topic of Chap. 4.

Along with the decay processes of Fig. 2.5a, one must also consider the corresponding inverse process, i.e. *pumping* from the paramagnon bath. Figure 2.5b shows the pumping processes. It is clear that the total decay rate for the probe paramagnon should be the difference

$$\Gamma_q(\omega) = \Gamma_q^{(d)}(\omega) - \Gamma_q^{(i)}(\omega) , \tag{2.9}$$

where Γ_q is the total width, $\Gamma_q^{(d)}$ is the decay width due to the processes in Fig. 2.5a and $\Gamma_q^{(i)}$ is the inverse width due to the processes in Fig. 2.5b. One can find a formal derivation, and further discussion, of (2.9) in Ref. [16]. The decay and inverse decay widths obtain a simple relation due to the condition of detailed balance [16, 17],

$$\Gamma_q^{(i)}(\omega) = e^{-\omega/T}\Gamma_q^{(d)}(\omega)$$
$$\Gamma_q(\omega) = (1 - e^{-\omega/T})\Gamma_q^{(d)}(\omega) \,. \tag{2.10}$$

Let us now write down an explicit expression for $\Gamma_q(\omega)$. A straightforward application of the Fermi Golden rule to the processes shown in Fig. 2.5 yields the following expression for the decay width

$$\begin{aligned}\Gamma_q(\omega) = (1 - e^{-\omega/T})\frac{16(2\pi)^6 \mathbb{S}\beta_0^2}{2\omega} \int \frac{d^3k_1}{2\omega_1(2\pi)^3}\frac{d^3k_2}{2\omega_2(2\pi)^3}\frac{d^3k_3}{2\omega_3(2\pi)^3} \\ \times [(1+n_1)(1+n_2)(1+n_3)\,\delta^{(4)}(q - k_1 - k_2 - k_3) \\ + 3n_1(1+n_2)(1+n_3)\,\delta^{(4)}(q + k_1 - k_2 - k_3) \\ + 3n_1 n_2(1+n_3)\,\delta^{(4)}(q + k_1 + k_2 - k_3) \\ + n_1 n_2 n_3\,\delta^{(4)}(q + k_1 + k_2 + k_3)] \,.\end{aligned} \tag{2.11}$$

Here

$$n_k = \frac{1}{e^{\omega_k/T} - 1} \tag{2.12}$$

is the paramagnon occupation number, and the four-dimensional δ-function accounts for the energy and momentum conservation of each process, and we have used the shorthand notation $\delta^{(4)}(q + k_1 + k_2 + k_3) = \delta(\omega_q + \omega_1 + \omega_2 + \omega_3)\delta^{(3)}(\boldsymbol{q} + \boldsymbol{k}_1 + \boldsymbol{k}_2 + \boldsymbol{k}_3)$. The constant pre-factor $\mathbb{S}$ is a combinatorial factor arising due to counting of the different possible paramagnon polarisations. Here we generalise to $O(N)$ field theory,

$$\mathbb{S} = 2(N+2). \tag{2.13}$$

For further details on such combinatorial factors see e.g. [18].

2.4.2 *Analysis of Quantum Disordered and Quantum Critical Regimes*

Having established a explicit expressions for the mass gaps, Néel temperature, and decay properties, we now wish to analyse the various proposed regimes of the phase diagram. Again, we are strictly considering the disordered phase. In nearly critical, two-dimensional quantum antiferromagnets it is well established that the phase diagram may be partitioned into three distinct regimes: the quantum disordered (QD), quantum critical (QC), and renormalised classical [19]. Moreover, it is also widely assumed, see e.g. Ref. [9], that nearly critical, three-dimensional quantum antiferromagnets may be similarly partitioned. In the disordered phase three regimes are commonly discussed: a quantum disordered (QD), quantum critical (QC),

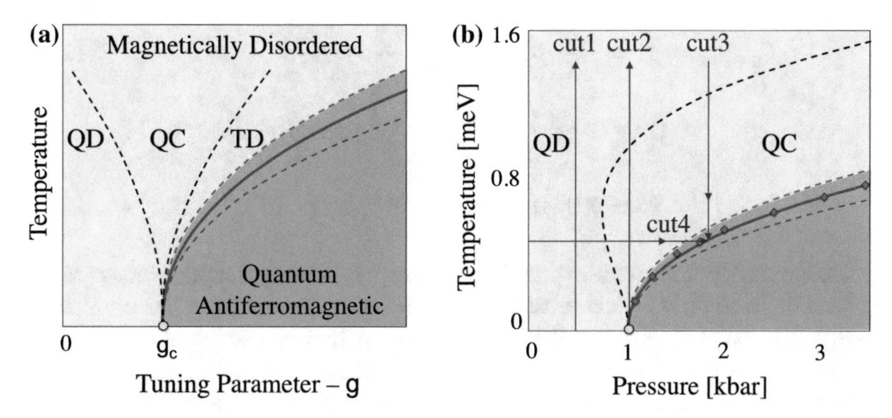

Fig. 2.6 The phase diagram and various regimes of a 3D quantum antiferromagnet. The dark red line is the Néel temperature. The light red band indicates the region of dimensional crossover or the classical critical regime. **a** Commonly accepted partition of the phase diagram into its various regimes. The dashed lines indicate crossovers between different regimes (not transitions). Here $g \sim p$ (pressure). **b** The phase diagram derived here, to be specific we use parameters of $TlCuCl_3$. The black dashed lines separate QD and QC regimes. The cuts: cut1, cut2, cut3, and cut4 are described in the text

and thermally disordered (TD). We illustrate such regimes in Fig. 2.6a. It is the purpose of this section to show that the inclusion of logarithmic corrections, via the running coupling constant, demands a significant change to this commonly held picture.

It will be convenient for the following technical discussion to work in terms or $\beta_\Lambda \equiv \alpha_\Lambda/(8\pi)$. Moreover, purely for the readers convenience we rewrite Eqs. (2.2) and (2.5) again here

$$\Delta^2 = \gamma^2(p_c - p)\left[\frac{\beta_\Lambda}{\beta_0}\right]^{\frac{N+2}{N+8}} + 8\pi(N+2)\beta_\Lambda \sum_{\mathbf{k}} \frac{1}{\omega_{\mathbf{k}}} \frac{1}{e^{\frac{\omega_{\mathbf{k}}}{T}} - 1}, \tag{2.14}$$

$$\beta_\Lambda = \frac{\beta_0}{1 + \frac{(N+8)\beta_0}{\pi}\ln(\Lambda_0/\Lambda)} \tag{2.15}$$

Again $\Lambda_0 = \max\{\Delta, T\}$, and $\omega_{\boldsymbol{k}} = \sqrt{k^2 + \Delta^2 + \Gamma^2}$.

To begin our analysis of the QD and QC regimes, we perform a simplifying approximation to the decay width expression (2.11); we consider just the Raman Raman processes shown Fig. 2.5a(b) and b(b). Such an approximation is certainly valid whenever $\Gamma \ll \Delta$, which constitutes most of QD and QC regimes. Within this approximation, we obtain

$$\Gamma_{q=0}(\omega=\Delta) = \frac{\pi\mathscr{S}}{2}\beta_\Lambda^2 T^3 \frac{1-e^{-\Delta/T}}{\Delta^2}\mathscr{I}\left(\frac{\Delta}{T}\right) ,$$
$$\mathscr{I}(y) = y\frac{6}{\pi^2}\int_y^\infty dx_1 \int_y^{x_1} dx_2\, n_{x_1}(1+n_{x_2})(1+n_{x_3}) ,$$
$$x_3 = y + x_1 - x_2 , \quad n_x = \frac{1}{e^x - 1}. \tag{2.16}$$

Importantly, within this expression we use the running coupling constant β_Λ instead of β_0 in (2.11). This substitution accounts for the one-loop RG corrections to Eq. (2.11), and implies that the paramagnons are non-interacting at the QCP.

We will also find it useful to calculate the Fermi golden rule $\Gamma_{q=0}(\omega)$ at arbitrary values of ω, i.e. off-mass-shell. As discussed, once the on-mass-shell restriction is lifted the spontaneous processes, Fig. 2.5a(a) and b(a), also contribute. In this case we find

$$\Gamma_{q=0}(\omega) = \frac{\pi\mathscr{S}}{2}\beta_\Lambda^2 T^3 \frac{1-e^{-\omega/T}}{\omega^2}\left\{\mathscr{I}_b\left(\frac{\omega}{T}\right) + \mathscr{I}_a\left(\frac{\omega}{T}\right)\right\} ,$$
$$\mathscr{I}_b(y) = y\frac{6}{\pi^2}\int_{\max\{y_0, 2y_0 - y\}}^\infty dx_1 \int_{y_0}^{y-y_0+x_1} dx_2$$
$$\times\, n_{x_1}(1+n_{x_2})(1+n_{x_3})F(x_1,x_2,x_3) ,$$
$$x_3 = y + x_1 - x_2 , \quad y_0 = \Delta/T ,$$
$$\mathscr{I}_a(y) = \theta(y - 3y_0)y\frac{2}{\pi^2}\int_{y_0}^\infty dx_1 \int_{y_0}^{y-y_0-x_1} dx_2$$
$$\times\, (1+n_{x_1})(1+n_{x_2})(1+n_{x_3})F(x_1,x_2,x_3) ,$$
$$x_3 = y - x_1 - x_2 , \quad y_0 = \Delta/T ,$$
$$F(x_1,x_2,x_3) = \begin{cases} 1, \text{ if } \ x_- \le x_3 \le x_+ \\ 0, \text{ otherwise} \end{cases}$$
$$x_- = \sqrt{\left(\sqrt{x_1^2 - y_0^2} - \sqrt{x_2^2 - y_0^2}\right)^2 + y_0^2} ,$$
$$x_+ = \sqrt{\left(\sqrt{x_1^2 - y_0^2} - \sqrt{x_2^2 - y_0^2}\right)^2 + y_0^2} . \tag{2.17}$$

When considering off-mass-shell energies ω, the running energy scale in β_Λ is set by $\Lambda = \max\{\sqrt{\omega^2 - q^2}, T\}$. It is easy to check that at $\omega = \Delta$, Eqs. (2.17) and (2.16) coincide.

2.4.2.1 Quantum Disordered Regime

To characterise the QD regime, consider the trace, denoted *cut1* in Fig. 2.6b, through the QD regime of the phase diagram. Deep within the QD regime and at low temperatures, such that $e^{-\Delta/T} \ll 1$, the gap evaluated via Eq. (2.14) receives negligible thermal corrections, and is essentially equal to its value at zero temperature. In this regime, we find that the decay width is given by

$$\frac{\Gamma_{q=0}(\omega = \Delta)}{\Delta} = \frac{3\mathbb{S}}{\pi}\beta_\Lambda^2 \frac{T^2}{\Delta^2} e^{-\Delta/T} \ll 1 , \tag{2.18}$$

which follows from a direct evaluation of (2.16).

2.4.2.2 Quantum Critical Regime

To characterise the QC regime, consider the trace, denoted *cut2* in Fig. 2.6b, which begins at the critical point $g = g_c$ and rises in temperature. Along cut2, the solution of Eq. (2.14) gives

$$\Delta = T\sqrt{\frac{2(N+2)\pi\beta_\Lambda}{3}}\,\Theta(\beta_\Lambda) . \tag{2.19}$$

We introduce the scaling function Θ, which is nonanalytic at $\beta \to 0$, and has the small β expansion $\Theta(\beta) = \left(1 - \sqrt{\frac{3(N+2)\beta}{2\pi}} + \cdots\right)$. This non analytic, cusp behaviour means that the expression (2.19) undergoes significant deviation from unity even at small values of β—the coupling constant. Figure 2.7 plots $\Theta(\beta)$ with $N = 3$.

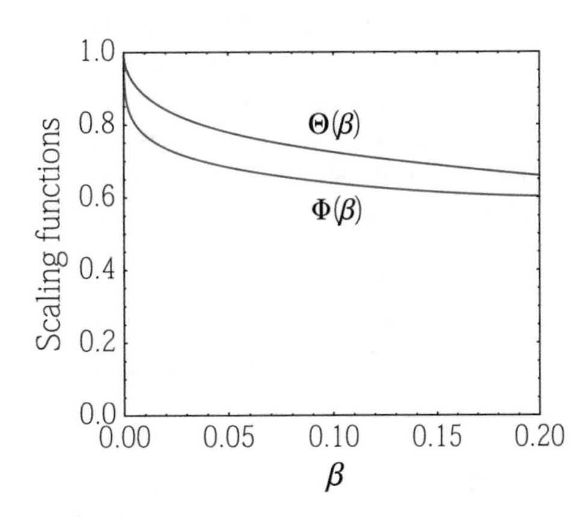

Fig. 2.7 Scaling functions $\Theta(\beta)$ and $\Phi(\beta)$ in Eqs. (2.19) and (2.20) for N = 3

We combine Eqs. (2.16) and (2.19) to find an appealing compact expression for the ration of the decay width to mass gap, appropriate to describe the cut2,

$$\frac{\Gamma_{q=0}(\omega = \Delta)}{\Delta} = \frac{3\mathbb{S}}{4(N+2)} \beta_\Lambda \, \Phi(\beta_\Lambda) \, . \tag{2.20}$$

We introduce a second scaling function Φ, which is also nonanalytic at $\beta \to 0$ and normalisation is chosen such that $\Phi(0) = 1$. See Fig. 2.7. Beginning at the QCP, where all energies vanish, and raising the temperature one may expect the observables Δ and Γ to scale linearly with temperature. Indeed, both Δ and Γ, Eqs. (2.19) and (2.20), scale linearly with temperature along the cut2. However, from the expressions (2.19) and (2.20) we also see a multiplicative logarithmic scaling. This is an interesting result, and offers yet another test of the infrared asymptotic freedom scenario. We also note that the logarithmic scaling for Δ and Γ are significantly different.

2.4.3 *Crossovers and Contours*

We have considered separately the QD and QC regimes, now we wish to understand the crossover curve between the two regimes. In the QD regime, $\Delta > T$, while in the QC regime, $\Delta < T$. Therefore one obtains the boundary curve by solving the implicit equation

$$\Delta(p, T) = T \, . \tag{2.21}$$

The black dashed line in Fig. 2.6b shows the crossover curve found via expressions (2.14) and (2.21). We immediately see that it does not follow the simple power scaling illustrated in Fig. 2.6a, and commonly referred to in the literature. The rather dramatic difference in the crossover curves of Fig. 2.6a and b is due to the running of the coupling constant. More physically, we state that this difference is due to the system being at its upper critical dimension, which implies that we must consider two energy scales: the infrared scale which is equal to temperature, and the ultraviolet scale which is determined by position of the Landau pole.

To be sure not to overstep the region of validity of the one-loop RG expressions presented in this section, we have to check that the running coupling remains smaller than unity $\beta \ll 1$. Obviously at the QCP $\beta \to 0$, but increasing T increases β. We evaluate β at the crossover $\Delta = T$ (the crossing point between the black crossover line and cut2 of Fig. 2.6b), and find it to be $\beta_c \approx 0.23$. Hence we see that coupling remains sufficiently small at the crossover to justify the perturbative results of this section.

To elucidate the physics of the so called thermal disordered (TD) regime, we consider the trace, denoted *cut3* in Fig. 2.6b. First we state that the ratio Δ/T monotonically decreases from $\Delta/T \geq 1$ above the QD-QC crossover, to $\Delta/T = 0$ at the Néel transition. Meanwhile the ratio Γ/Δ is monotonically increasing. On the

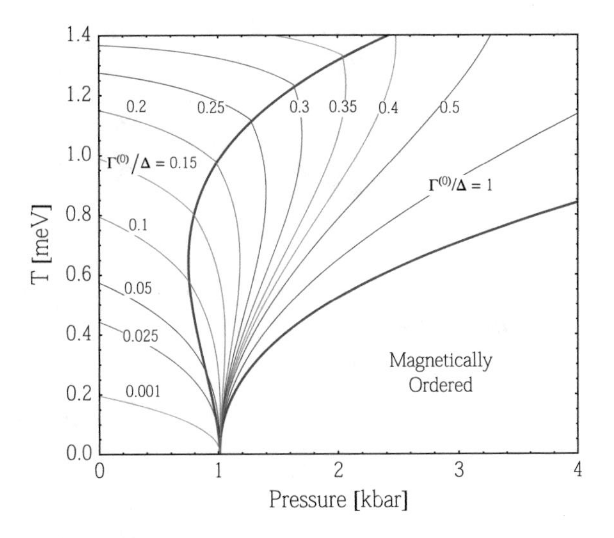

Fig. 2.8 Contours of $\Gamma^{(0)}/\Delta$ on the phase diagram of $TlCuCl_3$. The thick red line is the Néel temperature, the thick blue line is QD/QC crossover. All other curves are traces of constant $\Gamma^{(0)}/\Delta$

basis of this analysis, we do not observe any discernible fingerprints of a crossover to the TD regime. We suggest that this extra regime has very little qualitative distinction. However, once in the very near vicinity of the Néel temperature, the ratio Γ/Δ becomes approaches unity, and with it certainly signals a distinct regime. This regime change is known as *dimensional crossover*, and is indicated by the light red band in Fig. 2.6b. The region within the red band is often referred to as *classical critical*. We will explicitly develop a theoretical description for decay widths in such a regime in Chap. 4, which considers in detail cut3 and cut4. Moreover, in separate effort presented in Chap. 5 we develop a description of the static properties of the dimensional reduction.

To complete our discussion of the *crossovers and contours* of the magnetically disordered phase, Fig. 2.8 presents lines of constant Γ/Δ. The parameters (2.8) have been used. The cusps in the Γ/Δ contours occur at the QD-QC crossover line, i.e. when $\Delta = T$, we stress that such cusps are merely the result of the logarithmic RG where the argument is $\ln(\max\{\Delta, T\})$, and should not be mistaken for something physical.

2.5 Comparison with Experimental Data on $TlCuCl_3$

Having completed the theoretical discussion of the paramagnon gap, decay width and the associated regimes in the phase diagram, all that remains is to compare our results to available data on $TlCuCl_3$. Neutron scattering data is available in both the QD and QC regimes. We have already obtained the fitting parameters (2.8), and do not need any new parameters.

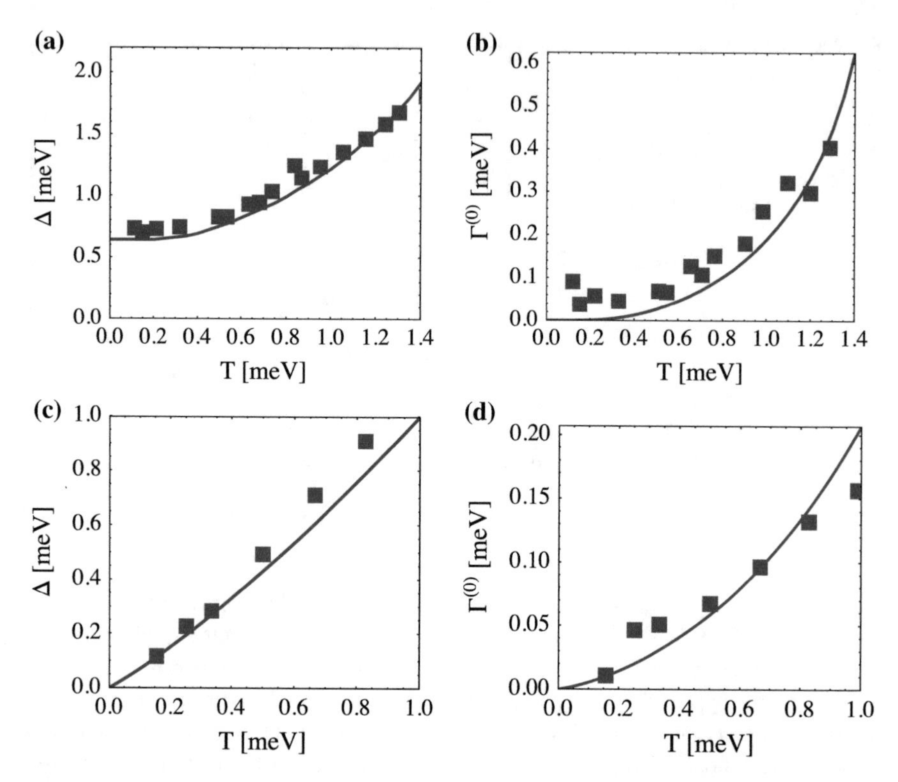

Fig. 2.9 **a** The gap and **b** the width in $TlCuCl_3$ along the cut1 in the phase digram Fig. 2.6b. We take the cut at $p = 0$ kbar. Squares represent experimental data Ref. [20] and the theory is shown by lines. **c** The gap and **b** the width in $TlCuCl_3$ along the critical cut2, $p = p_c$, in the phase digram Fig. 2.6b. Squares represent experimental data Ref. [6] and the theory is shown by lines

First, we consider an analogous trace to cut1 in Fig. 2.6b, except we demand that it begins at $p = 0$, such that it corresponds to experimental data. Figure 2.9a and b shows the paramagnon gap (2.14) and width (2.14) along this trace. Our theoretical results are shown by solid lines and the squares represent the experimental data of Ref. [20]. We have not used Eq. (2.18), since it is valid only up to $\exp(-\Delta/T) \ll 1$. The agreement between experiment and theory for both the gap and the width is remarkable.

Next we consider the trace analogous to cut2 in Fig. 2.6b, i.e. the quantum critical regime. Figure 2.9c and d shows the paramagnon gap (2.19)and width (2.20) along this trace. Our theoretical results are shown by solid lines and the squares represent the experimental data of Ref. [20]. Once again, our theoretical results are in remarkable agreement with the experimental data. We emphasise that our calculations of the gaps and widths entering Fig. 2.9 had no adjustable fitting parameters; the small set of parameters were already determined in Sect. 2.3.

2.6 Discussion and Conclusion

Asymptotic freedom is inherent to the mathematics of relativistic quantum field theories at the upper critical dimension. The present work has demonstrated that this mathematical intrigue finds a physical realisation in the setting of 3D quantum antiferromagnets. First we establish the existence of this remarkable property in $TlCuCl_3$. We then go on to explore the implications of asymptotic freedom on various observables. Explicitly, we show how the inclusion of the running coupling plus renormalization due to statistical fluctuations provides a remarkably accurate description of the mass gap and decay width of magnons in $TlCuCl_3$. Our results also provide a natural description of the boundary/crossover to the so called *quantum critical regime*—we determine the boundary to be qualitatively different from the commonly accepted picture.

It is important to emphasise that it was the remarkable experimental control of quantum antiferromagnet $TlCuCl_3$ that has allowed the present work to identify—for the first time—the logarithmic decay of the coupling constant.

References

1. Gross DJ, Wilczek F (1973) Ultraviolet behavior of non-abelian gauge theories. Phys Rev Lett 30:1343–1346
2. Politzer HD (1973) Reliable perturbative results for strong interactions? Phys Rev Lett 30:1346–1349
3. Landau LD, Abrikosov AA, Khalatnikov IM (1954) The removal of infinities in quantum electrodynamics. Dokl Akad Nauk SSSR 95:497; An asymptotic expression of the electron Green function in quantum electrodynamics 95:773; An asymptotic expression of the photon Green function in quantum electrodynamics 95:1177
4. Tanaka H, Goto K, Fujisawa M, Ono T, Uwatoko Y (2003) Magnetic ordering under high pressure in the quantum spin system $TlCuCl_3$. Phys B Condens Matter:329–333, (Part 2):697–698. Proceedings of the 23rd international conference on low temperature physics
5. Nohadani O, Wessel S, Haas S (2005) Quantum phase transitions in coupled dimer compounds. Phys Rev B 72:024440
6. Merchant P, Normand B, Kramer KW, Boehm M, McMorrow DF, Rüegg C (2014) Quantum and classical criticality in a dimerized quantum antiferromagnet. Nat Phys 10(5):373–379
7. Rüegg C, Furrer A, Sheptyakov D, Strässle T, Krämer KW, Güdel H-U, Mélési L (2004) Pressure-induced quantum phase transition in the spin-liquid $TlCuCl_3$. Phys Rev Lett 93:257201
8. Rüegg C, Normand B, Matsumoto M, Furrer A, McMorrow DF, Krämer KW, Güdel HU, Gvasaliya SN, Mutka H, Boehm M (2008) Quantum magnets under pressure: controlling elementary excitations in $TlCuCl_3$. Phys Rev Lett 100:205701
9. Sachdev S (2011) Quantum phase transitions. Cambridge University Press
10. Kulik Y, Sushkov OP (2011) Width of the longitudinal magnon in the vicinity of the O(3) quantum critical point. Phys Rev B 84:134418
11. Goldstone J, Salam A, Weinberg S (1962) Broken symmetries. Phys Rev 127:965–970
12. Sachdev S (2009) Exotic phases and quantum phase transitions: model systems and experiments. arXiv:0901.4103
13. Zinn-Justin J (2002) Quantum field theory and critical phenomena. International series of monographs on physics. Clarendon Press

14. Yan QQ, Normand B, Sandvik AW, Meng ZY (2015) Multiplicative logarithmic corrections to quantum criticality in three-dimensional dimerized antiferromagnets. Phys Rev B 92:214401
15. Note that in the ordered phase spontaneous decay of a magnon is generally allowed. For an analysis at zero temperature, we draw the readers attention to the work of Refs. [83,140,141]
16. Weldon HA (1983) Simple rules for discontinuities in finite-temperature field theory. Phys Rev D 28:2007–2015
17. Donoghue JF, Holstein BR (1983) Renormalization and radiative corrections at finite temperature. Phys Rev D 28:340–348
18. Peskin ME, Schroeder DV (1995) An introduction to quantum field theory. Westview
19. Chakravarty S, Halperin BI, Nelson DR (1989) Two-dimensional quantum Heisenberg antiferromagnet at low temperatures. Phys Rev B 39:2344–2371
20. Rüegg C, Normand B, Matsumoto M, Niedermayer C, Furrer A, Krämer KW, Güdel H-U, Bourges P, Sidis Y, Mutka H (2005) Quantum statistics of interacting dimer spin systems. Phys Rev Lett 95:267201

Chapter 3
Unifying Static and Dynamic Properties in 3D Quantum Antiferromagnets

Abstract Quantum Monte Carlo offers an unbiased means to study the static and dynamic properties of quantum critical systems, while quantum field theory provides direct analytical results in terms of the quasiparticle excitations. We study three dimensional, critical quantum antiferromagnets performing a combined analysis by means of quantum field theory calculations and quantum Monte Carlo data. Explicitly, we analyse the order parameter (staggered magnetisation), Néel temperature, quasiparticle gaps, as well as the susceptibilities in the scalar and vector channels. We connect the two approaches by deriving descriptions of the quantum Monte Carlo observables in terms of the quasiparticle excitations of the field theory, which reduces the number of fitting parameters. Agreement is remarkable, and constitutes a thorough test of perturbative $O(3)$ quantum field theory. We outline future avenues of research the present work opens up.

3.1 Introduction

Quantum field theory (QFT) and quantum Monte Carlo (QMC) are two indispensable methods to study critical phenomena in magnetic quantum systems—each offering a different perspective into the characteristic phenomena of the system. Quantum field theory comes with a set of fundamental parameters which completely determine all observables and, in particular, how they scale with detuning, δg, from the critical point g_c (see Fig. 3.1). The parameters are phenomenological and must be determined by fitting to experimental or numerical data. On the other hand, the QMC simulations considered here are performed using a lattice Hamiltonian and all observables are expressed in terms of the detuning parameter δg as well as a set of arbitrary fitting parameters; the fitting parameters of any observable being unrelated to those of another.

Recent QMC studies [1–3] have demonstrated excellent agreement with the predicted scaling behaviour of observables [4, 5], and has only become possible with recent developments in computational techniques. However, the shortcoming of such

H. Scammell, *Interplay of Quantum and Statistical Fluctuations in Critical Quantum Matter*, Springer Theses,
https://doi.org/10.1007/978-3-319-97532-0_3

a treatment is that each observable is detached from the others; information relating observables is lost in QMCs arbitrary fitting parameters. This is unsatisfying since one generally expects, and finds from QFT, that observables are intimately linked; e.g. scaling behaviour of the Néel transition temperature can be determined by the properties of the critical excitations. Our primary motivation is to derive the relations between all such quantities, and provide a complete mapping between the QMC and QFT observables.

In the vicinity of the magnetic critical point the observables of interest are associated with the broken or unbroken $O(3)$ symmetry. This group accounts for the relevant, i.e. critical, degrees of freedom. As described in Chap. 2, the symmetric (disordered) phase hosts three gapped triplon modes, while in the symmetry broken phase, a preferred direction is established and is associated with an order parameter or staggered magnetisation, see Fig. 3.1a. The amplitude oscillation of the order parameter is the gapped Higgs mode and the directional oscillations are gapless Goldstone modes. In three spatial dimensions, order survives up to a non-zero Néel temperature. An illustration of the phase diagram and some observables is presented in Fig. 3.1a.

Again, the aim of the present work is to connect all such observables via a description in terms of a set of five QFT parameters $\{c, g_c, \gamma, \alpha_0, \Lambda_0\}$ to be described shortly. Explicitly, the parameters $\{c, g_c\}$ will be fixed to the values obtained in simulations, and the remaining three are to be determined using best fit to QMC data.

The remainder of the chapter is organised as follows: Sect. 3.2 provides a description of the lattice Hamiltonian used in QMC. Section 3.3 details the meanfield quantum field theory and single-loop RG corrections. In Sect. 3.4 we apply the analytic QFT formulae to fit the QMC data [1, 2]. Section 3.5 provides a detailed analysis of the vector and scalar response functions, and offers a self-contained treatment. We use the derived parameters to analyse the Higgs decay linewidth obtained from the vector and scalar response functions in [2]. In Sect. 3.6 we derive approximate values of the best-fit parameters from bond-operator theory. Here we also explain the non-universal relationship between m_s and φ_c. Section 3.7 discusses the findings and suggests future research avenues.

3.2 Model and Methods

As a representative 3D dimerised lattice with an unfrustrated geometry, we choose to study the double cubic model shown in Fig. 3.1b. This system consists of two interpenetrating cubic lattices with the same antiferromagnetic interaction strength, J, connected pairwise by another antiferromagnetic interaction, J'. The QPT occurs when the coupling ratio $g = J'/J$ is increased, changing the ground state from a Néel-ordered phase of finite staggered magnetisation to a dimer-singlet (quantum disordered) phase, as illustrated in Fig. 3.1a. The Hamiltonian reads,

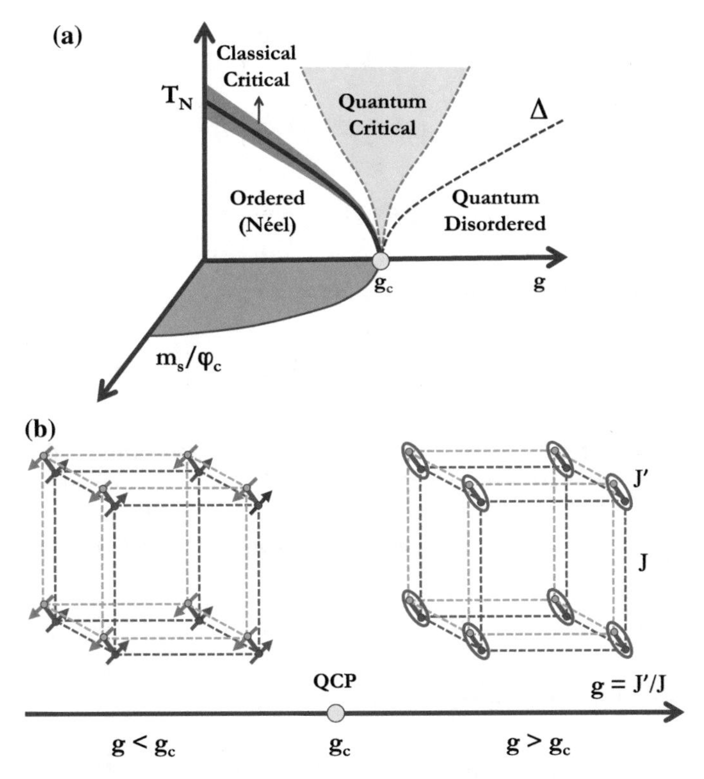

Fig. 3.1 **a** Schematic quantum critical phase diagram for the Heisenberg model on the double cubic lattice. Staggered magnetisation (red) and Néel temperature (blue) and triplon gap (dashed) all vanish at the QCP $g = g_c$. **b** Dimerised lattice of $S = 1/2$ spins in the $3D$ double cubic geometry. Sites of the red and light green cubic lattices are connected pairwise by dimer bonds; J' and J are antiferromagnetic Heisenberg interactions on and between the dimer units, respectively, and their ratio, $g = J'/J$, controls the QPT from a Néel ordered phase (left) to a quantum disordered dimer-singlet phase (right), with the QCP occurring at the critical ratio g_c

$$H = J \sum_{<i,j>} \{\mathbf{S}_l^i \cdot \mathbf{S}_l^j + \mathbf{S}_r^i \cdot \mathbf{S}_r^j\} + J' \sum_i \mathbf{S}_l^i \cdot \mathbf{S}_r^i, \tag{3.1}$$

where subscripts $\{l, r\}$ denote the left and right position on the dimer.

3.2.1 *Observables*

The observables of interest in the QMC simulations are the triplon and Higgs excitation gaps, and the staggered magnetisation, all at zero temperature. QMC also determines the Néel temperature. The zero temperature observables can be cast in the following generic form [1–5]

$$\Delta_t(g) = a_1|g - g_c|^{\nu_1} \ln\left[\frac{|g - g_c|}{b_1}\right]^{\beta_1}, \tag{3.2}$$

$$\Delta_H(g) = a_2|g - g_c|^{\nu_2} \ln\left[\frac{|g - g_c|}{b_2}\right]^{\beta_2}, \tag{3.3}$$

$$m_s(g) = a_3|g - g_c|^{\nu_3} \ln\left[\frac{|g - g_c|}{b_3}\right]^{\beta_3}, \tag{3.4}$$

while the Néel temperature is written [1, 5]

$$T_N(g) = a_4|g - g_c|^{\nu_4} \ln\left[\frac{|g - g_c|}{b_4}\right]^{\beta_4}. \tag{3.5}$$

The exponents $\{\nu_i, \beta_i\}$ have received a great deal of attention, and are known from scaling hypotheses and general quantum field theoretic arguments; $\nu_i = 1/2$, $\beta_1 = \beta_2 = -\frac{1}{2}(N+2)/(N+8)$ and $\beta_3 = \beta_4 = 3/(N+8)$ [4]. It is the relationship between all coefficients $\{a_i, b_i\}$ that remains unknown from QMC analysis. We will derive such relations.

3.3 Quantum Field Theory

The quantum phase transition (QPT) between ordered and disordered phases is described by the effective field theory with the following Lagrangian [6],

$$\mathcal{L} = \frac{1}{2}\partial_\mu \vec{\varphi}\partial^\mu \vec{\varphi} - \frac{1}{2}m_0^2 \vec{\varphi}^{\,2} - \frac{1}{4}\alpha_0[\vec{\varphi}^{\,2}]^2. \tag{3.6}$$

The vector field $\vec{\varphi}$ describes staggered magnetisation, and index μ enumerates time and three coordinates, and $\partial_\mu = (\partial_t, c\nabla)$ where c is the (magnon) spin wave velocity. The QPT results from tuning the mass term, m_0^2, for which we take the linear expansion $m_0^2(\delta g) = \gamma^2 (g - g_c)/g_c$, where $\gamma^2 > 0$ is a coefficient and g is the quantum tuning parameter. Varying g leads to two distinct phases: (i) for $g > g_c$ we have $m_0^2 > 0$, and the classical expectation value of the field is zero $\varphi_c^2 = 0$. (ii) For $g < g_c$ we have $m_0^2 < 0$, and the field obtains a non-zero classical expectation value $\varphi_c^2 = \frac{|m_0^2|}{\alpha_0}$. As per Chap. 2 one easily recovers the known relation for the bare (unrenormalised) parameters; Higgs gap/triplon gap$= \sqrt{2}$, explicitly $\Delta_t(\delta g) = m_0(\delta g)$ and $\Delta_H(\delta g) = \sqrt{2}|m_0(\delta g)|$.

The above analysis does not account for quantum or thermal fluctuations. All fluctuations considered in the present chapter originate from the vertex and self-energy diagrams shown in Fig. 3.2. Appendix A provides a full derivation such corrections.

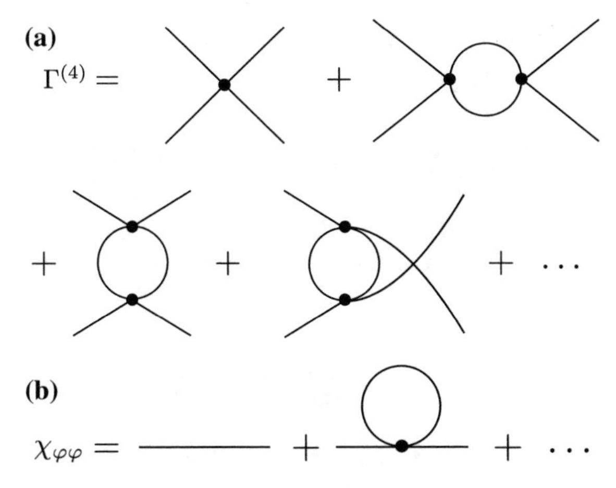

Fig. 3.2 Diagrammatic expansions for **a** the four-point vertex, $\Gamma^{(4)}$, and **b** the response function, $\chi_{\varphi\varphi}$, shown for the quantum disordered phase ($g > g_c$). Solid lines denote the free propagation, governed by the first two terms of Eq. (3.6), of the field φ, which here corresponds to triplon propagation. The vertex marked by the solid circle represents the bare interaction, the third term of Eq. (3.6), whose coefficient, α, is the perturbative parameter. The one-loop corrections to $\Gamma^{(4)}$ and $\chi_{\varphi\varphi}$ are equivalent to retaining next-to-leading-order terms in α. For the expansion of $\Gamma^{(4)}$ this implies α^2 terms, which are contained in the three distinct loop diagrams (the Mandelstam s, t, and u channels) in panel (**a**). For the expansion of $\chi_{\varphi\varphi}$ this is the order-α loop diagram in panel (**b**), to which we refer as the self-energy, Σ

3.4 Results

For comparison with QMC, we have the following four observables as derived in one-loop renormalisation group, Appendix A. We rewrite the analytic form of the zero temperature excitation gaps and order parameter in a more convenient form

$$\Delta_t^2(\delta g) = \gamma^2|\delta g|\left[\frac{\alpha_\Delta}{\alpha_0}\right]^{\frac{N+2}{N+8}} = \frac{\gamma^2}{g_c}\left(\frac{16\pi^2}{(N+8)\alpha_0}\right)^{\frac{N+2}{N+8}} |g-g_c|\left|\ln\left(|g-g_c|/\tilde{b}_1\right)\right|^{-\frac{N+2}{N+8}}, \quad (3.7)$$

$$\Delta_H^2(\delta g) = 2\gamma^2|\delta g|\left[\frac{\alpha_\Delta}{\alpha_0}\right]^{\frac{N+2}{N+8}} = 2\frac{\gamma^2}{g_c}\left(\frac{16\pi^2}{(N+8)\alpha_0}\right)^{\frac{N+2}{N+8}} |g-g_c|\left|\ln\left(|g-g_c|/\tilde{b}_2\right)\right|^{-\frac{N+2}{N+8}}, \quad (3.8)$$

$$\varphi_c^2(\delta g) = \frac{\gamma^2|\delta g|}{\alpha_0}\left[\frac{\alpha_0}{\alpha_\Delta}\right]^{\frac{6}{N+8}} = \frac{\gamma^2}{\alpha_0 g_c}\left(\frac{16\pi^2}{(N+8)\alpha_0}\right)^{\frac{-6}{N+8}} |g-g_c|\left|\ln\left(|g-g_c|/\tilde{b}_3\right)\right|^{\frac{6}{N+8}}, \quad (3.9)$$

and similarly for the Néel temperature

$$T_N(\delta g)^2 = \frac{12\gamma^2|\delta g|c^3}{(N+2)\alpha_0}\left[\frac{\alpha_0}{\alpha_{T_N}}\right]^{\frac{6}{N+8}}$$
$$= \frac{12\gamma^2 c^3}{(N+2)\alpha_0 g_c}\left(\frac{16\pi^2}{(N+8)\alpha_0}\right)^{\frac{-6}{N+8}} |g-g_c| \left|\ln\left(|g-g_c|/\tilde{b}_4\right)\right|^{\frac{6}{N+8}} . \quad (3.10)$$

Here $\delta g = (g - g_c)/g_c$, $N = 3$, $c = 2.365$ and $g_c = 4.83704$ are extracted from QMC simulations [1]. The logarithmic scale dependence of the running coupling constant is

$$\alpha_\Lambda = \frac{\alpha_0}{1 + \frac{(N+8)\alpha_0}{8\pi^2}\ln(\Lambda_0/\Lambda)}, \quad (3.11)$$

where $\Lambda = \max\{\Delta_t, \Delta_H/\sqrt{2}, T\}$ is set by the largest energy scale, and Λ_0 is the normalisation point. Note, we take running scale to be $\Delta_H/\sqrt{2} = |\Delta_t|$ when plotting $\Delta_H(\delta g)$ and $m_s(\delta g)$. Exact knowledge of coefficients within the logarithms requires two-loop RG. Our choice of Δ_t and $|\Delta_t|$ for the disordered and ordered phases, respectively, provides scaling in δg symmetric with respect to the QCP $\delta g = 0$.

On the right-most equality in each Eqs. (3.7, 3.8, 3.9, 3.10) shows a re-parametrisation to give a scaling form identical to Eqs. (3.2, 3.3, 3.4, 3.5). The constants $\tilde{b}_i$ are

$$\tilde{b}_1 = \tilde{b}_2 = \tilde{b}_3 = \frac{g_c\Lambda_0^2}{\gamma^2} e^{\frac{16\pi^2}{(N+8)\alpha_0}},$$
$$\tilde{b}_4 = \frac{(N+2)\alpha_0 g_c \Lambda_0^2}{12c^3\gamma^2} e^{\frac{16\pi^2}{(N+8)\alpha_0}}. \quad (3.12)$$

This is achieved through reparametrising α_Λ to show explicit dependence on detuning δg

$$\alpha_\Lambda(\delta g) = \frac{16\pi^2}{(N+8)}\left|\ln\left(|g-g_c|/\tilde{b}_i\right)\right|^{-\frac{N+2}{N+8}}, \quad (3.13)$$

where $\tilde{b}_i$ corresponds to $\tilde{b}_1$ for $T = 0$, and to $\tilde{b}_4$ along the Néel temperature curve. The running coupling is uniquely determined by Eq. (3.11) as a function of the energy scale ratio Λ_0/Λ. However, reparametrising in terms of detuning $|g - g_0|$, Eq. (3.13), we must include the constant $\tilde{b}_i$ to account for the different possible dependences of Λ_0/Λ on $|g - g_c|$. The parametrisation (3.13) serves two purposes, first it allows for simple conversion between QFT running coupling constant *language*, and the logarithmic scaling forms used widely in condensed matter literature, see e.g. Ref. [4]. Second, it explicitly shows that the five fundamental parameters of the QFT uniquely determine the functional form of all four observables above, including coefficients. Put differently, the free parameters in QMC $\{a_i, b_i\}$ and exponents $\{\nu_i, \beta_i\}$ are now directly obtained. Next we provide numerical values of such parameters.

3.4.1 Best-Fit Parameters

The best-fit parameters are found to be

$$\Lambda_0 = 0.915J; \quad \alpha_0/(8\pi c^3) = 0.175; \quad \gamma = 3.95J, \tag{3.14}$$

while $c = 2.365$ and $g_c = 4.83704$, are extracted from QMC simulations [1]. It is important to note that the choice of Λ_0 is arbitrary, and that it affects the value chosen for α_0, since $\alpha_0 \equiv \alpha_{\Lambda_0}$.

The relation between m_s and φ_c is found by assuming a form

$$\varphi_c = \Upsilon m_s \tag{3.15}$$

with $\Upsilon =$constant. Fitting Eq. (3.9) to QMC data [1] we obtain $\Upsilon = 0.65$. The proportionality constant Υ does not appear in the quantum field theory, however an approximate value for Υ is derived in Sect. 3.6.2 by appealing to a bond-operator technique.

We are now ready to demonstrate the agreement between QMC and QFT. The Fig. 3.3a, b and c show results of Eqs. (3.7, 3.8, 3.9, 3.10), with parameters (3.14), plotted on log-log axes. Agreement is remarkable, and clearly demonstrates that QFT, with a single set of parameters, is capable of quantitatively describing the static and dynamic observables. This procedure demonstrates, to a high precision, the validity of the theoretical predictions of the $O(3)$ QFT, and constitutes our main result. Moreover, the results shown in Fig. 3.3a and b demonstrate excellent agreement over a large range of detuning from the quantum critical point.

3.5 Higgs Decay Linewidth

In this section we analyse the Higgs decay linewidth of the three dimensional dimerised QAF. In QFT, QMC and experiment the linewidth is extracted from an appropriate response function. Experimentally, neutron scattering is a successful technique to probe Higgs or triplon modes, for example studies of $TlCuCl_3$ [7–9]. Neutron scattering constitutes a vector probe and so provides access to a vector susceptibility. In QFT and QMC, one is free to use vector or scalar probes to extract information about the system. Recent QMC studies [2, 3] have performed state-of-the-art numerical analytic continuation of imaginary time Greens functions, thus allowing a numerical study of vector and scalar response functions in three dimensional dimerised QAFs. We will now analyse the vector and scalar response functions using QFT, and the parameters derived in Sect. 3.4. The following analysis is restricted to the ordered phase, where spontaneous decay of the Higgs mode is possible. In the disordered phase, spontaneous decay of triplons is forbidden due to a lack of available phase space, see e.g. Ref. [10].

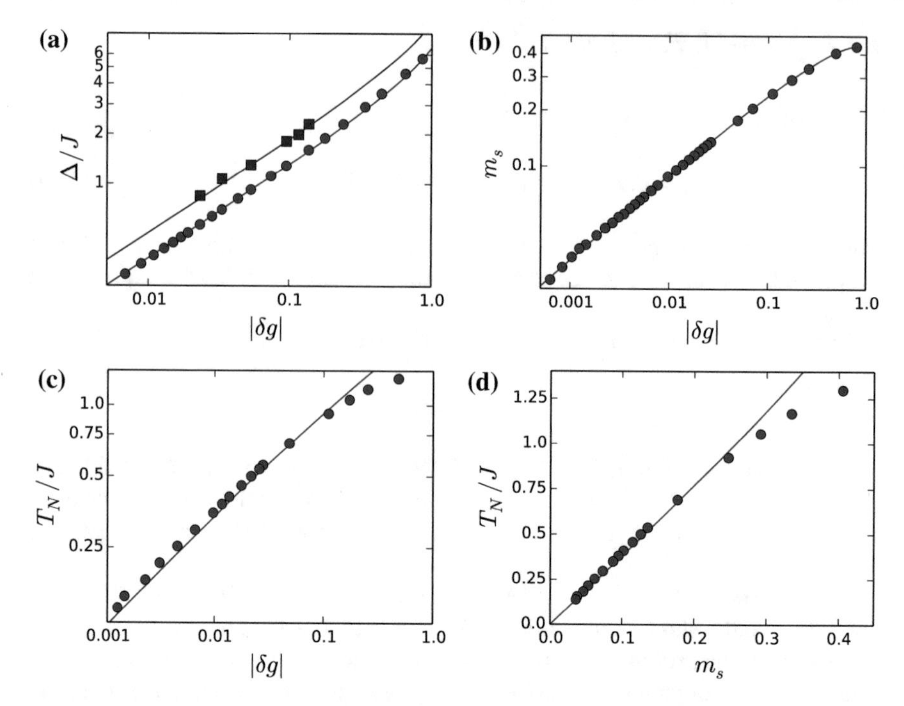

Fig. 3.3 **a** $\Delta = \{\Delta_t, \Delta_H\}$ as a function of tuning parameter δg. QMC data [1] shown by blue circles for Δ_t, and indigo squares for Δ_H, QFT fits Eqs. (3.7) and (3.8) are given by blue and indigo lines. **b** m_s as a function of tuning parameter δg on log-log axes. QMC data [1] shown by blue markers m_s, QFT fit (3.9) is given by blue line. **c** T_N as a function of tuning parameter δg. QMC data [1] shown by blue markers, QFT fit Eq. (3.10) given by blue line. **d** T_N versus m_s, blue points corresponds to QMC data [1], blue line corresponds to QFT fits derived from Eqs. (3.9) and (3.10) with $m_s = \Upsilon^{-1}\varphi_c$

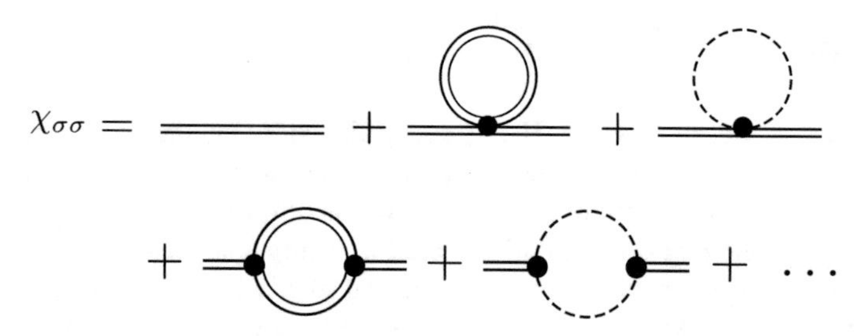

Fig. 3.4 Diagrammatic representation of the loop contributions to $\chi^{(v)}_{\sigma\sigma}$, to frst-order in α. Double and dashed lines represent Higgs and Goldstone propagators, respectively

3.5.1 Vector Response

We now consider the vector susceptibility, $\chi_v = \langle \vec{\varphi}(p)\vec{\varphi}(0)\rangle$. In the ordered phase we write $\vec{\varphi} = (\varphi_c + \sigma, \vec{\pi})$, and decompose the vector susceptibility into the Higgs and Goldstone components,

$$\begin{aligned}\chi_v &= \langle \sigma(p)\sigma(0)\rangle + (N-1)\langle \pi(p)\pi(0)\rangle \\ &= \chi_{\sigma\sigma} + (N-1)\chi_{\pi\pi}\end{aligned} \tag{3.16}$$

written in this form, the vector response essentially corresponds to an unpolarised probe and as such is averaged over all components. Since the QMC simulations are performed on finite size lattices, spin rotation symmetry remains unbroken and response functions are rotationally averaged analogously to Eq. (3.16). Note, cross components of the Higgs field σ and order parameter φ_c do not contribute $\chi_{\sigma\varphi_c} = 0$.

We now include the frst-order in α corrections to the susceptibility. The goldstone component of the susceptibility $\chi_{\pi\pi}$ does not receive any corrections from the loop diagrams, this is a direct result of the Goldstone theorem explicitly demonstrated in Appendix A. The Higgs component receives loop corrections, as shown in Fig. 3.4, and the real part of the loop corrections has been explicitly treated in Appendix A. In all following equations, the Higgs gap Δ_H represents the single-loop renormalised value. It remains then to evaluate the imaginary contribution of the loop-corrections to the Higgs susceptibility. The first two loop-diagrams in the right hand side of Fig. 3.4 have purely real contribution and are already accounted for within Δ_H. The second two have an imaginary contribution and are denoted Π_H(p), $\Pi_G(p)$ for the polarisation loop with two Higgs internal lines, and two Goldstone internal lines, respectively, and p is the external four momentum. Again, the real part of such polarisation loops have already been taken into account, so we are just evaluating the imaginary parts, denoted $\Pi''_H(p)$ and $\Pi''_G(p)$. To this order, the susceptibilities are

$$\chi_{\sigma\sigma}(p) = \frac{1}{p^2 - \Delta_H^2 + \frac{i}{2}\alpha_\Lambda \Delta_H^2 \left[\Pi''_G(p) + \Pi''_H(p)\right]} \tag{3.17}$$

$$\chi_{\pi\pi}(p) = \frac{1}{p^2 + i0}. \tag{3.18}$$

where $\Pi''_G(p)$ and $\Pi''_H(p)$ are the imaginary parts of the polarisation loops with two Goldstone, and two Higgs propagators, respectively, and are given by standard loop-integrals, see e.g. [11, 12]

$$\Pi''_G(p) = \frac{(N-1)}{8\pi}\theta(p^2), \tag{3.19}$$

$$\Pi''_H(p) = \frac{9}{8\pi}\frac{\sqrt{p^2 - 4\Delta_H^2}}{p}\theta(p^2 - 4\Delta_H^2). \tag{3.20}$$

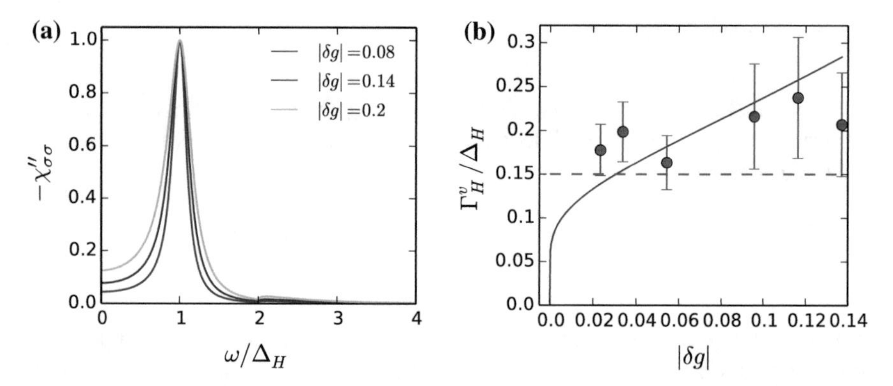

Fig. 3.5 **a** Imaginary part of the vector response function, given by $-\chi''_{\sigma\sigma}(\omega)$, shown as a function of ω/Δ_H at $\boldsymbol{p} = \boldsymbol{0}$ and normalized to its maximum value. The curves correspond to different values, $|\delta g|$, of the coupling ratio relative to the QCP. **b** Ratio Γ^v_H/Δ_H of the Higgs line width, as determined from the vector response function, to its gap, shown as a function of $|\delta g|$. The solid line is the QFT result obtained from Eq. (3.21). The dashed line is the ratio extracted from QMC data by averaging over $|\delta g|$ and extrapolating in system size ($L \to \infty$) [2]. The points are obtained from the QMC data for systems of sizes $L = 14$ and 16 at the different values of $|\delta g|$ for which simulations were performed

Here $p^2 = \omega^2 - \boldsymbol{p}^2$ is the external four momentum, and θ is the Heaviside theta function. The vector of momentum is measured from the antiferromagnetic ordering vector, $\boldsymbol{Q}$.

Spectral functions come from the imaginary part of the susceptibility. We are primarily interested in the spectral linewidth of the Higgs mode at zero spatial momentum $\boldsymbol{p} = \boldsymbol{0}$. To this end, we plot the spectral function $-\chi''_{\sigma\sigma}(\omega, \boldsymbol{0})$ in Fig. 3.5a. It takes on a Lorentzian shape, with Full-width at half-maximum linewidth (decay width)

$$\Gamma^v_H(|\delta g|) = \frac{\alpha_\Lambda}{8\pi c^3}\Delta_H(|\delta g|) = \frac{\alpha_0 \Delta_H(|\delta g|)}{8\pi c^3\left[1 + \frac{(N+8)}{8\pi^2 c^3}\alpha_0 \ln(\sqrt{2}\Lambda_0/\Delta_H)\right]}. \tag{3.21}$$

This linewidth exactly corresponds to width calculated in Ref. [13] using Fermi golden rule. And physically it corresponds to the process of a Higgs mode spontaneously decaying into two Goldstone modes. The process of the Higgs mode decaying into two Higgs modes, i.e. $\Pi''_H(\omega, \boldsymbol{0})$, has a threshold at $\omega = 2\Delta_H$, and is found not to contribute to the linewidth Γ_H.

Importantly, the Higgs decay width (in vector channel) as given by Eq. (3.21) is completely determined by the fundamental parameters of the QFT. Hence, having determined the best fit parameters (3.14), we can now predict the corresponding decay linewidth of the Higgs mode. We plot the results of Eq. (3.21) as a function of $|\delta g|$, the blue line in Fig. 3.5b. The decay width reduces to zero logarithmically in accord with the asymptotic freedom of the QCP, $|\delta g| = 0$. The data in red is from

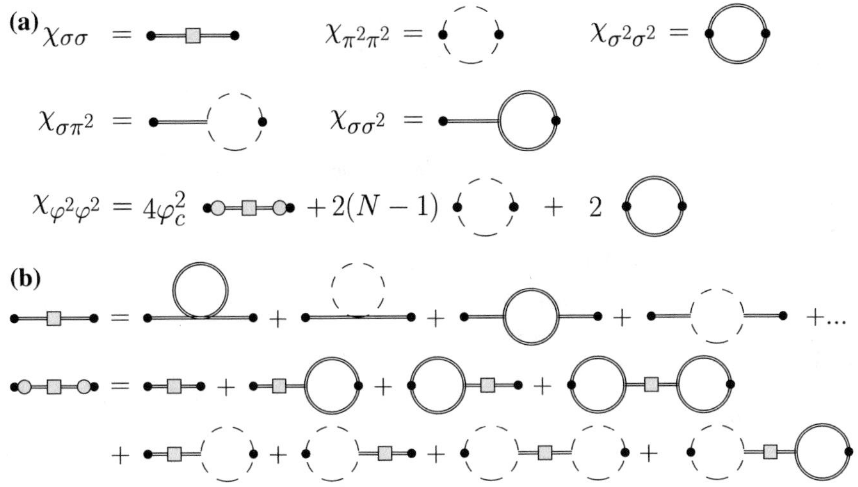

Fig. 3.6 **a** Diagrammatic representation of the contributions to $\chi_{\varphi^2\varphi^2}$, with notation defined in **b**. **b** Pole and vertex renormalisation. The diagrammatic sub-series to frst-order in α

the QMC simulations, Ref. [2]. We note that QMC data is not sufficiently accurate to discern logarithmic corrections. Even so, the overlap between QFT and QMC, as shown in Fig. 3.5b, is convincing and warrants future studies.

3.5.2 Scalar Response

We now consider the scalar susceptibility, $\chi_{\varphi^2\varphi^2} = \langle \vec{\varphi}^2(p)\ \vec{\varphi}^2(0)\rangle$. Again, in the ordered phase we write $\vec{\varphi} = (\varphi_c + \sigma, \vec{\pi})$ allowing for a decompose the scalar susceptibility into the Higgs and Goldstone components,

$$\chi_{\varphi^2\varphi^2} = 4\varphi_c^2\chi_{\sigma\sigma} + 4\varphi_c(\chi_{\sigma\pi^2} + \chi_{\sigma\sigma^2}) + \chi_{\pi^2\pi^2} + \chi_{\sigma^2\sigma^2} + 2\chi_{\sigma^2\pi^2}. \tag{3.22}$$

For the scalar susceptibility, $\chi_{\varphi^2\varphi^2}$, we will consider contributions at order $\mathcal{O}(\alpha^0)$. We ignore the final term $\chi_{\sigma^2\pi^2}$ since it only contributes at order $\mathcal{O}(\alpha^1)$, see Ref. [11, 12] for further details.

Figure 3.6 provides the explicit sub-series to the desired order $\mathcal{O}(\alpha^0)$ contributing to $\chi_{\varphi^2\varphi^2}$. Evaluation of this series gives Eq. (3.23). We explicitly show tadpole contributions in the top line of Fig. 3.6b, and how they are to be incorporated in all other summations. Tadpoles must be properly accounted to provide the correct critical indices in Eqs. (3.8), (3.9), and (3.10). Importantly, after resummation of the top two lines in Fig. 3.6 one obtains the $\chi_{\sigma\sigma}$ (double line), which has identical pole structure to the vector response; the real and imaginary parts of the Higgs pole are identical. Performing the diagrammatic resummation outlined in Fig. 3.6, we obtain

$$\chi_{\varphi^2\varphi^2}(p) = \frac{2\Delta_H^2}{\alpha_\Lambda} \frac{\left\{1 - i\alpha_\Lambda \left[\Pi_G''(p) + \frac{1}{6}\Pi_H''(p)\right]\right\}^2}{p^2 - \Delta_H^2 + \frac{i}{2}\alpha_\Lambda \Delta_H^2 \left[\Pi_G''(p) + \Pi_H''(p)\right]} - i\left[\Pi_G''(p) + \frac{1}{9}\Pi_H''(p)\right], \tag{3.23}$$

where $\Pi_G''(p)$ and $\Pi_H''(p)$ are given by Eqs. (3.19, 3.20).

There are two important aspects to note in the scalar susceptibility: First, the numerator in the first term on RHS of Eq. (3.23) contains a phase factor (imaginary part), which comes from vertex renormalisation, see Fig. 3.6. Second, there are non-resonant pole contributions given by the second term on the RHS of Eq. (3.23). The addition of the phase factor and non-resonant pole terms results in a destructive interference in the emission channel of two low energy Goldstone modes. The interference acts to suppress the imaginary part of $\chi_{\varphi^2\varphi^2}(p) \sim p^4$ at $p \to 0$, which is a statement of the Adler theorem. To explicitly show this, we rearrange the expression (3.23) and present the p^4 dependence of the imaginary part of $\chi_{\varphi^2\varphi^2}$ at $p^2 < 4\Delta_H^2$,

$$\chi_{\varphi^2\varphi^2}'' = \frac{-p^4\Pi_G''(p)}{(p^2 - \Delta_H^2)^2 + (\frac{1}{2}\alpha\Delta_H^2\Pi_G''(p))^2}\,. \tag{3.24}$$

In this expression we ignore contributions due to $\Pi_H(p)$, since they do not contribute to the imaginary part for $p^2 < 4\Delta_H^2$.

In the limit of large momentum $p^2 \gg \Delta_H^2$, the non-resonant pole terms in (3.23) dominate. These terms correspond to the background scattering which, from Eqs. (3.19) and (3.20), have large-p asymptotic form

$$\Pi_G''(p) + \frac{1}{9}\Pi_H''(p^2 \gg \Delta_H^2) \to \frac{3}{8\pi}. \tag{3.25}$$

Taking $p = (\omega, \mathbf{0})$, Eq. (3.25) accounts for the spectral weight of the large-ω tail in Fig. 3.7.

We now make some general remarks about the results of this section and the QMC results of Refs. [2, 3]. The line shape, Fig. 3.7a, is a Fano resonance with additional interference resulting in ω^4 suppression at low energy. This asymmetric shape compares well with the recent QMC results [2, 3]. However, in the present work we have paid special attention to the logarithmic corrections and have found that their inclusion prevents any 'universal data collapse', which has been approximately observed in [2, 3]. The data in red is from the QMC simulations, Ref. [2]. Moreover, in [2] the averaged linewidth to gap ratio is found to be $\Gamma_H/\Delta_H \approx 0.43$, while from the present analysis we find that Γ_H/Δ_H is essentially identical to that found in the vector response, Eq. (3.21), and is shown in Fig. 3.7b. Disagreement in the case of the scalar response function requires further studies. For example, it is possible that the error bars from QMC have a broadening effect on the spectral function obtained from the stochastic analytic continuation. One could therefore perform further studies to test how significantly the statistical error bars affect the shape of certain types of spectral functions.

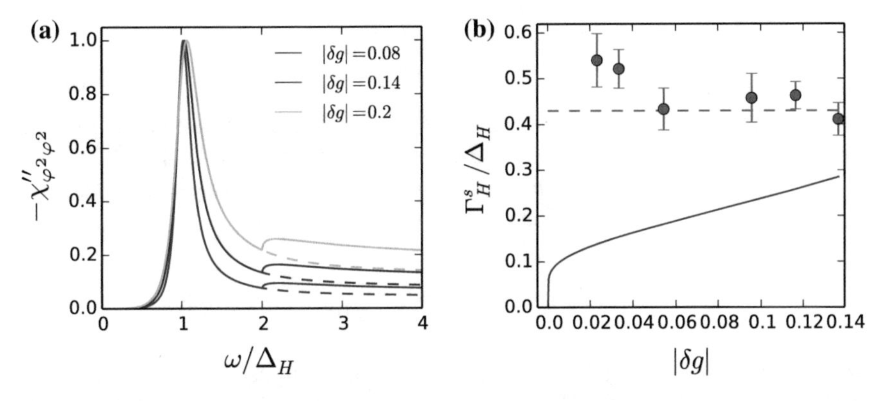

Fig. 3.7 **a** Imaginary part of the scalar response function, $-\chi''_{\varphi^2\varphi^2}(\omega)$, shown as a function of ω/Δ_H at $\boldsymbol{p} = \boldsymbol{0}$ and normalized to its maximum value. The curves are evaluated from Eq. (3.23) and correspond to different values, $|\delta g|$, of the coupling ratio relative to the QCP. Dashed lines show the results obtained from Eq. (3.23) but neglecting the two-Higgs contribution [Eq. (3.20)]. **b** Ratio Γ^s_H/Δ_H of the Higgs line width, as determined from the scalar response function, to its gap, shown as a function of $|\delta g|$. The solid line is the QFT result obtained from Eqs. (3.23) and (3.21). The dashed line is the ratio extracted from QMC data by averaging over $|\delta g|$ and extrapolating in system size ($L \to \infty$) [2]. The points are obtained from the QMC data for systems of sizes $L = 14$ and 16 at the different values of $|\delta g|$ for which simulations were performed

3.6 Derivation of Parameters

In this section we motivate the values of best-fit parameters (3.14) and Υ from Eq. (3.15) by appealing directly to the lattice Hamiltonian (3.1). Explicitly, we derive parameters γ, c, g_c and Υ in terms of J/J'.

3.6.1 Gap, Velocity, and Critical Point

A widely used technique to analyse spin-dimerised quantum magnets is to employ a bond-operator representation [14] of the spin-1/2 operators of the Hamiltonian. Applying such a technique to (3.1), we can estimate the coefficient γ in the QFT gap Δ_t, Eq. (3.7), from the corresponding gap found using bond operators Δ_{BO}. Straightforward analysis presented in Appendix B yields $\Delta_{\text{BO}} = \sqrt{A_Q^2 - B_Q^2}$, where A_q and B_q are defined in Eq. (B.6), and $\boldsymbol{Q} = (\pi, \pi, \pi)$ is the antiferromagnetic ordering vector.

The idea is to equate the two gaps at the normalisation point Λ_0, i.e. $\Delta_t(\Lambda_0) = \Delta_{\text{BO}}(\Lambda_0) = \Lambda_0$, where the last equality defines our normalisation point. The normalisation point $\Lambda_0 = 0.915J$ is chosen in this way and we remind the reader that this choice is essentially arbitrary. Equating the gaps provides

$$\gamma^2 = \frac{\Delta_{\mathrm{BO}}^2(\Lambda_0)}{|\delta g(\Lambda_0)|} \ . \tag{3.26}$$

A simple evaluation of Eq. (3.7) gives $|\delta g(\Lambda_0)| \approx 0.056$ and we obtain the estimate $\gamma = 3.88J$, which compares well with $\gamma = 3.95J$ obtained in Eq. (3.14).

The critical point is found in this approximation by setting $\Delta_{\mathrm{BO}} = 0$, and we find the spin wave velocity at the critical point to be

$$g_c = 4.96, \qquad c = \lim_{q \to Q} \frac{\Omega_q(g_c)}{|q - Q|} = 2.28, \tag{3.27}$$

where $\Omega_q(g)$ is the Bogolyubov spectrum given in Appendix B. These values compare reasonably well with those extracted from QMC [1], $g_c = 4.83704$ and $c = 2.365$.

3.6.2 Relating m_s and φ_c

From QFT alone one cannot directly obtain the staggered magnetisation, m_s. Instead QFT provides the order parameter φ_c. To obtain the relation between m_s and φ_c, we appeal to the triplon bond-operator $\vec{t}$ and find the proportionality factor relating it to $\vec{\varphi}$

$$\vec{\varphi} = Z^{-1} \frac{1}{2} (\vec{t}^{\,\dagger} + \vec{t}) \tag{3.28}$$

Employing the Bogolyubov representation,

$$\varphi(x) = \sum_k \frac{1}{\sqrt{2\Omega_k}} \left\{ \beta_k e^{ikx} + \beta_k^\dagger e^{-ikx} \right\}, \tag{3.29}$$

$$t(x) = \sum_k \left\{ u_k \beta_k - v_{-k} \beta_{-k}^\dagger \right\} e^{ikx}, \tag{3.30}$$

$$\approx \sum_k \sqrt{\frac{A_k}{2\Omega_k}} \left\{ \beta_k e^{ikx} + \beta_k^\dagger e^{-ikx} \right\}, \tag{3.31}$$

where u_k and v_k are usual Bogoliubov coefficients, defined in Appendix B. In the vicinity of the QCP the most important contribution comes from the low energy excitations with $\boldsymbol{q} \sim \boldsymbol{Q}$. We therefore approximate the proportionality factor as

$$Z = \frac{1}{\sqrt{A_Q}}. \tag{3.32}$$

The staggered magnetisation of an antiferromagnet with sublattice A and B, reads as follows

$$m_s^z = \frac{1}{N}\langle S_A^z - S_B^z\rangle \tag{3.33}$$

where, $S^{A,B} = \sum_i^{N'} S_i^{A,B}$, and $N' = \frac{1}{2}N$. Performing the bond-operator transformation [14]

$$S_i^{A,B} = \frac{1}{2}(\pm s_i^\dagger t_{i,\alpha} \pm t_{i,\alpha}^\dagger s_i - i\epsilon_{\alpha,\beta,\gamma} t_{i,\beta}^\dagger t_{i,\gamma}) \tag{3.34}$$

where the $s_i^\dagger/s_i$ are singlet creation/annihilation operators on bond site i. We replace $s_i^\dagger/s_i$ with the condensate value $\langle s\rangle = \langle s^\dagger\rangle = \bar{s} = 0.97$, and details of calculation are left for the Appendix B. Therefore,

$$\begin{aligned} m_s^z &= \frac{1}{2N}\langle\sum_{i\in A}^{N'}(s_i^\dagger t_{i,z} + t_{i,z}^\dagger s_i) + \sum_{i\in B}^{N'}(s_i^\dagger t_{i,z} + t_{i,z}^\dagger s_i)\rangle \\ &= \frac{\bar{s}}{2}\langle t_z + t_z^\dagger\rangle = \bar{s}Z\langle\varphi_z\rangle \end{aligned} \tag{3.35}$$

$$\Upsilon = \frac{\varphi_c}{m_s^z} = \sqrt{\frac{1}{\bar{s}^2 A_Q}} = 0.62. \tag{3.36}$$

where $\langle\varphi_z\rangle = \varphi_c$. This value compares well with that used for fitting $\Upsilon = 0.65$ in Fig. 3.3b and d. Furthermore, using the relation derived in Ref. [15]

$$T_N = c^{3/2}\sqrt{\frac{12}{5}}\varphi_c \tag{3.37}$$

$$= \Upsilon c^{3/2}\sqrt{\frac{12}{5}}m_s. \tag{3.38}$$

we therefore expect (up to logarithmic corrections) that T_N is proportional to m_s. This has been explicitly considered in [1, 16, 17], and is clear from Fig. 3.3d.

3.7 Discussion and Conclusion

At a pragmatic level, the present work offers a means for direct comparison between QMC and QFT, and explicitly derives the parameters relating the J–J' Hamiltonian (3.1) on the double cubic lattice geometry to the QFT (3.6). We now discuss future research avenues that could utilise and benefit from the present results.

First, there remains three unresolved issues from the present analysis: (i) The rather abrupt disagreement of the fits to $T_N(|\delta g|)$ for $|\delta g| \gtrsim 0.2$; does this imply the limit of quantum critical scaling or, instead, is it an issue with numerics at larger temperatures? (ii) Although the line shape of the scalar spectral function shows excellent agreement for QFT and QMC, the width of the scalar spectral function found in the two approaches disagrees by more than a factor of two. We believe such significant disagreement cannot be assigned to the error margins. (iii) In the vector channel, the Higgs linewidth shows reasonable agreement between the two approaches. However, the current QMC data has insufficient accuracy to discern logarithmic corrections to this quantity. It would be desirable for future numerical studies to focus on the logarithmic dependence of the Higgs linewidth, which is expected since the theory becomes asymptotically free at the quantum critical point. All three questions require further QMC studies to resolve.

Second, the present work considered the zero temperature behaviour of the gaps and order parameter. However, non-zero temperature behaviour of these observables and are completely determined by the results of the present work i.e. an analysis of the non-zero temperature properties would require no new fitting parameters. Going to non-zero temperatures induces many exotic phenomena not present at zero temperature. In particular, one generally expects finite temperature crossovers into regions of the phase diagram known as classical critical and quantum critical, see Fig. 3.1a. Such crossovers are thought to have a significant influence on the scaling behaviour of the gaps and order parameter. Extending the QMC to non-zero temperatures and performing a similar analysis to that provided here would therefore provide a quantitative examination of such a scenario.

Non-zero temperature also generates additional scattering channels for quasiparticles, i.e. scattering from the heat bath, to be discussed in Chap. 4 next. This can have many physical implications. For example, triplons in the disordered phase at zero temperature have zero decay width (infinite lifetimes), however, through heat bath scattering the triplons can acquire a substantial decay width. This scenario has been considered in three dimensional quantum antiferromagnet $TlCuCl_3$ [9], and discussed analytically in [10]. A corresponding QMC study of triplon decay at non-zero temperatures has yet to be performed and is certainly an interesting possibility for future work.

Finally, we comment on the possibility to extend the present results to include the influence of an applied, static magnetic field, B. The addition of the magnetic field provides another tuning handle to generate symmetry breaking and, importantly, the corresponding critical observables would follow from this work without need for additional fitting parameters. There are a number of interesting predictions for critical scaling behaviour in the case of magnetic field, to be discussed in Chaps. 7 and 8. It is straightforward to extend the results present work, without introducing new parameters, to account for an applied magnetic field and hence the present results could be directly applied to test such predictions.

In summary, this chapter provides a detailed mapping between QFT results derived in terms of quasiparticles and QMC data obtained directly in terms of the spin Hamiltonian. The purpose is to offer insight into the connection between the static and dynamical properties of critical systems. Moreover, a description of the observables in terms of quantum field theory allows the number of unknown parameters to be reduced and serves as a basis for future quantitative tests of the low energy effective quantum field theory against unbiased quantum Monte Caro.

References

1. Qin YQ, Normand B, Sandvik AW, Meng ZY (2015) Multiplicative logarithmic corrections to quantum criticality in three-dimensional dimerized antiferromagnets. Phys Rev B 92:214401
2. Qin YQ, Normand B, Sandvik AW, Meng ZY (2017) Amplitude mode in three-dimensional dimerized antiferromagnets. Phys Rev Lett 118:147207
3. Lohöfer M, Wessel S (2017) Excitation-gap scaling near quantum critical three-dimensional antiferromagnets. Phys Rev Lett 118:147206
4. Zinn-Justin J (2002) Quantum field theory and critical phenomena. International series of monographs on physics. Clarendon Press
5. Scammell HD, Sushkov OP (2015) Asymptotic freedom in quantum magnets. Phys Rev B 92:220401 Dec
6. Sachdev S (2011) Quantum phase transitions. Cambridge University Press
7. Rüegg C, Normand B, Matsumoto M, Furrer A, McMorrow DF, Krämer KW, Güdel HU, Gvasaliya SN, Mutka H, Boehm M (2008) Quantum magnets under pressure: controlling elementary excitations in $TlCuCl_3$. Phys Rev Lett 100:205701
8. Rüegg C, Furrer A, Sheptyakov D, Strässle T, Krämer KW, Güdel H-U, Mélési L (2004) Pressure-induced quantum phase transition in the spin-liquid $TlCuCl_3$. Phys Rev Lett 93:257201
9. Merchant P, Normand B, Kramer KW, Boehm M, McMorrow DF, Rüegg C (2014) Quantum and classical criticality in a dimerized quantum antiferromagnet. Nat Phys 10(5):373–379
10. Scammell HD, Sushkov OP (2017) Nonequilibrium quantum mechanics: a "hot quantum soup" of paramagnons. Phys Rev B 95:024420
11. Podolsky D, Auerbach A, Arovas DP (2011) Visibility of the amplitude (Higgs) mode in condensed matter. Phys Rev B 84:174522
12. Katan YT, Podolsky D (2015) Spectral function of the Higgs mode in $4-\epsilon$ dimensions. Phys Rev B 91:075132
13. Kulik Y, Sushkov OP (2011) Width of the longitudinal magnon in the vicinity of the O(3) quantum critical point. Phys Rev B 84:134418
14. Sachdev S, Bhatt RN (1990) Bond-operator representation of quantum spins: mean-field theory of frustrated quantum Heisenberg antiferromagnets. Phys Rev B 41:9323–9329
15. Oitmaa J, Kulik Y, Sushkov OP (2012) Universal finite-temperature properties of a three-dimensional quantum antiferromagnet in the vicinity of a quantum critical point. Phys Rev B 85:144431
16. Jin S, Sandvik AW (2012) Universal Néel temperature in three-dimensional quantum antiferromagnets. Phys Rev B 85:020409
17. Tan D-R, Jiang F-J (2017) Universal scaling of Néel temperature, staggered magnetization density, and spin-wave velocity of three-dimensional disordered and clean quantum antiferromagnets. Phys Rev B 95:054435

Chapter 4
A Nonperturbative Theory of Paramagnon Decay

Abstract We consider the dynamics of paramagnons in 3D quantum antiferromagnets at nonzero temperature and in the vicinity of the quantum critical point. Upon approach to the phase transition, the heat bath causes infrared divergences in the paramagnon decay width calculated using standard perturbative approaches. To describe this regime we develop a new finite frequency, finite temperature technique for a nonlinear quantum field theory—the 'golden rule of quantum kinetics'. The formulation is generic and applicable to any three dimensional quantum antiferromagnet in the vicinity of a quantum critical point. We obtain all results in the generic $O(N)$ quantum field theory. Specifically we apply our results to $TlCuCl_3$ (where we take $N = 3$) and find compelling agreement with experimental data.

4.1 Introduction

The interplay between quantum and statistical fluctuations in quantum systems offers many exciting challenges to theory. In particular, developing appropriate techniques to describe (quasi)particles in a hot and dense medium is a challenge of fundamental importance to many areas of physics ranging from condensed matter, to plasma, nuclear, and particle physics. In this case, the challenge arises because standard perturbative treatments of quasiparticles in hot dense mediums become plagued by infrared divergences and are hence unreliable. In this chapter we concentrate on lifetimes of quasiparticles or, more generally, on the line-shapes of quasiparticle spectral functions.

Our primary motivation is to develop and present a relatively simple technique that cures the enigmas of perturbative approaches, and offers reliable results in the physically interesting regimes of hot, dense mediums. To this end, our developed technique possesses two key properties: (i) it regulates the infrared behaviour via a resummation of medium effects i.e. the self-consistent inclusion of line-shapes, and (ii) allows one to handle the calculation of non-equilibrium responses at finite temperature.

Physically, the problem we consider was stimulated by the observation of magnetic excitations (paramagnons) in the magnetically disordered phase of the three dimen-

H. Scammell, *Interplay of Quantum and Statistical Fluctuations in Critical Quantum Matter*, Springer Theses,
https://doi.org/10.1007/978-3-319-97532-0_4

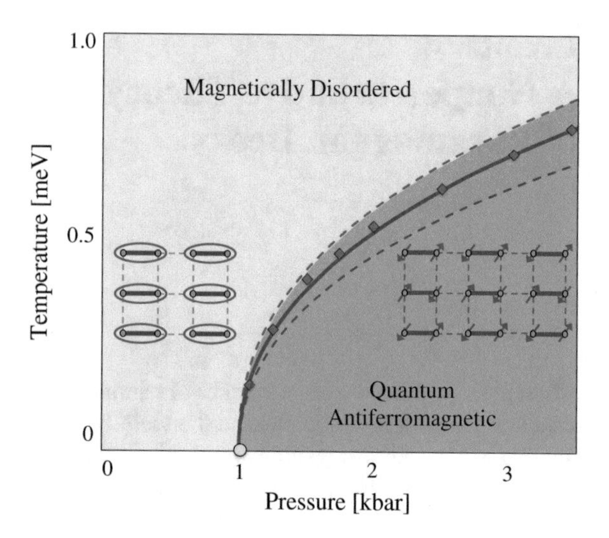

Fig. 4.1 The phase diagram of $TlCuCl_3$. The Néel temperature curve (solid red line) separates magnetically ordered and magnetically disordered phases. The quantum critical point (yellow dot) is at $p = p_c = 1.01$ kbar. Points show experimental data from Ref. [3]. The light red band around the Néel curve indicates the region of dimensional crossover

sional, dimerised quantum antiferromagnet $TlCuCl_3$ [1]. The pressure-temperature phase diagram of the compound is shown in Fig. (4.1). The data in Ref. [1] is taken over a range of pressures and temperatures that go beyond the regime of validity of standard perturbation theory, and hence provides an exiting opportunity to apply and test our developed technique.

It is important to note that the developed techniques are generic and applicable to all systems of this kind, i.e. symmetric phases described by $O(N)$ field theories. For example, they are applicable to the electroweak phase transition in cosmology, to the wide class of spin dimerised magnetic models [2], and to $O(2)$ superfluids or superconductors in the vicinity of their quantum critical points.

The remainder of this chapter is organised as follows: Sect. 4.2 introduces the necessary mathematical background. Section 4.3 discusses the inconsistency of the usual perturbative Fermi golden rule, and introduces our proposed *golden rule of quantum kinetics*, which simultaneously incorporates decay and heat bath scattering processes in a self-consistent formalism. A general mathematical analysis of the golden rule of quantum kinetics, without reference to any particular system, is given in Sect. 4.4. Finally in Sect. 4.5 we apply our technique to the specific compound $TlCuCl_3$ and directly compare our results with inelastic neutron scattering data.

4.2 General Considerations

We employ a quantum field theoretic description of the system and its excitations [4, 5]

$$\mathscr{L} = \frac{1}{2}\partial_\mu \vec{\varphi} \partial^\mu \vec{\varphi} - \frac{1}{2} m_0^2 \vec{\varphi}^2 - \frac{1}{4}\alpha_0 \vec{\varphi}^4, \tag{4.1}$$

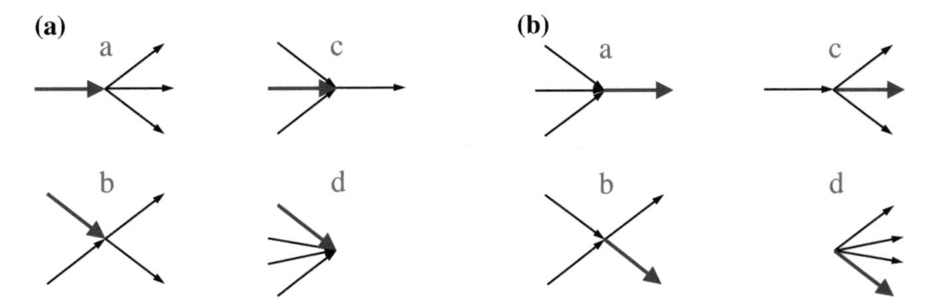

Fig. 4.2 **a** Decay diagrams for a paramagnon. The thick blue line represents the probe paramagnon and thin black lines represent the heat bath paramagnons. **b** Diagrams corresponding to pumping (inverse processes) to the paramagnon state. The thick blue line represents the probe paramagnon and thin black lines represent the heat bath paramagnons

where $\vec{\varphi} = (\varphi_1, \varphi_2, \varphi_3)$ is a three component real vector field which describes the three polarisations of the paramagnons. The partial derivates are defined in $3+1$ dimensional Minkowski spacetime $\partial_\mu = (\partial_0, c\bar{\nabla})$, and the paramagnon speed is set equal to unity, $c = 1$. We take the linear expansion $m_0^2 = \gamma^2(p_c - p)$, where γ is a coefficient and p represents, for example, an external pressure. We find it convenient to present equations using the rescaled coupling constant,

$$\beta = \frac{\alpha}{8\pi} \, . \tag{4.2}$$

Throughout we work in the regime whereby $\beta \ll 1$, such that perturbation theory is reliable. We apply the renormalization group (RG) to obtain the scale dependence of the coupling constant and the mass; they become energy, momentum, and temperature dependent, $\beta_0 \to \beta_q, m_0^2 \to m_q^2$. We have checked that $\beta \ll 1$ throughout the majority of phase diagram Fig. 4.1, and hence the results for the mass and running coupling derived within one-loop RG (in previous chapters and Ref. [6]), are valid throughout all regions of interest to the present chapter.

The focus of the present chapter is the decay width and spectral function of paramagnons in the magnetically disordered phase, Fig. 4.1. Unlike the running coupling and mass, the width and spectral function cannot be obtained within standard one-loop perturbation theory.

A schematic of the paramagnon decay channels, in the presence of a heat bath, is present in Fig. 4.2. They can be understood as the imaginary part of the *sunset diagram* presented in Fig. 4.3.

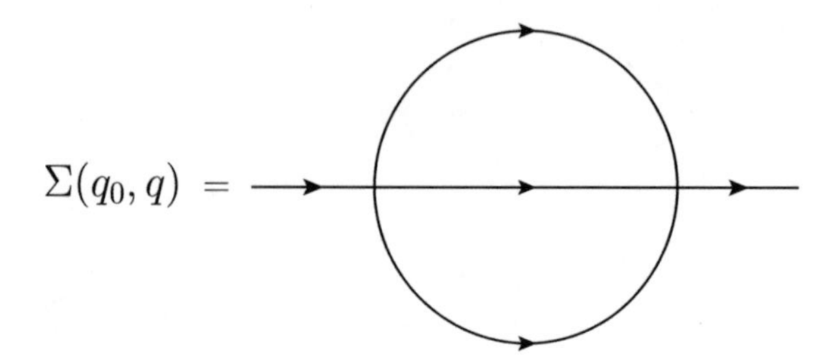

Fig. 4.3 The *sunset* self-energy diagram

4.2.1 The Spectral Function

The key mathematical object under consideration is the retarded Greens function of the paramagnon. To allow for a coherent presentation, we outline some essential properties of the retarded Greens function $G^R(\omega, q)$. Consider the noninteracting field theory, i.e. setting $\beta = 0$ in the Lagrangian (4.1),

$$\mathscr{L} = \frac{1}{2}\partial_\mu \vec{\varphi} \partial^\mu \vec{\varphi} - \frac{1}{2} m_0^2 \vec{\varphi}^2 \ . \tag{4.3}$$

This Lagrangian permits an exact Greens function

$$G^R(\omega, \mathbf{q}) = \frac{1}{2\omega_q} \left\{ \frac{1}{\omega - \omega_q + i0} - \frac{1}{\omega + \omega_q + i0} \right\} ,$$
$$\omega_q = \sqrt{q^2 + m_0^2} , \tag{4.4}$$

a conclusion which holds at both zero and nonzero temperatures, as long as interaction terms are absent from Eq. (4.3). We also immediately deduce the simple symmetry properties of G^R; the real and imaginary parts of G^R are even and odd functions of ω, respectively. Importantly, these generic symmetry properties carry over to the case of non-zero interaction $\beta \neq 0$.

Once interactions are taken into account, exactly obtaining the Greens function is a delicate, if not impossible, task. However, we can still make use of the following exact relation

$$-\frac{1}{\pi} Im\ G^R(\omega, \boldsymbol{q}) = (1 - e^{-\omega/T}) S_q(\omega) , \tag{4.5}$$

which is valid at arbitrary interaction and arbitrary temperature.

Taking into account the interaction term $\alpha_0 \vec{\varphi}^4/4$ in Eq. (4.1), leads to loop corrections to the Greens function. Put in another way, we must take into account paramagnon self-energy $\Sigma_q(\omega)$.

Including the real part of the self-energy was the topic of Chap. 2. Under one-loop renormalisation group (RG) the real part leads to the replacement $m_0^2 \to m_q^2$ in Eq. (4.4). We further use notation $m_q \equiv \Delta$ to denote the renormalised mass at nonzero temperature, such that the dispersion is given by

$$\omega_q = \sqrt{q^2 + \Delta^2}\,, \tag{4.6}$$

and Δ depends on momentum and temperature. Below we take ω_q as given by Eq. (4.6). Following from Chap. 2 and [6], the gap equation in the paramagnetic phase is given by

$$\Delta^2 = \gamma^2(p_c - p)\left[\frac{\beta_\Lambda}{\beta_0}\right]^{\frac{N+2}{N+8}} + 8\pi(N+2)\beta_\Lambda \sum_{\mathbf{k}} \frac{1}{\omega_k} \frac{1}{e^{\frac{\omega_k}{T}} - 1}\,. \tag{4.7}$$

Here have generalised to an N component theory, and β_Λ is the running coupling constant

$$\beta_\Lambda = \frac{\beta_0}{1 + \frac{(N+8)\beta_0}{\pi}\ln(\Lambda_0/\Lambda)}\,. \tag{4.8}$$

The normalisation point Λ_0 is set by $\Lambda = \max\{\Delta, T\}$.

The imaginary part of the self-energy describes broadening (decay)

$$\Gamma_q(\omega) = -\frac{Im\Sigma_q(\omega)}{\omega} \tag{4.9}$$

$$G^R(\omega, \mathbf{q}) = \frac{1}{\omega^2 - \omega_q^2 - \Sigma_q(\omega)} \to \frac{1}{\omega^2 - \omega_q^2 + i\omega\Gamma_q(\omega)}\,.$$

There are two points to note, (i) generally Γ_q depends on ω and hence the line shape can be significantly different from that of a simple Lorentzian, and (ii) $\Gamma_q(\omega)$ is an even function of ω since $Im\Sigma_q(\omega)$ is an odd function. The spectral function corresponding to (4.9) immediately follows from Eq. (4.5),

$$S_q(\omega) = \frac{1}{\pi(1 - e^{-\frac{\omega}{T}})}\left\{\frac{\omega\Gamma_q}{[\omega^2 - \omega_q^2]^2 + \omega^2\Gamma_q^2}\right\}. \tag{4.10}$$

In this chapter we are explicitly interested in the sunset diagram Fig. 4.3. Following the calculation presented in Chap. 2, we find that the width function, as defined in Eq. (4.9), is given by

$$\begin{aligned}
\Gamma_q(\omega) = (1 - e^{-\omega/T})&\frac{16(2\pi)^6 \mathbb{S}\beta_\Lambda^2}{2\omega} \int \frac{d^3k_1}{2\omega_1(2\pi)^3}\frac{d^3k_2}{2\omega_2(2\pi)^3}\frac{d^3k_3}{2\omega_3(2\pi)^3} \\
&\times [(1+n_1)(1+n_2)(1+n_3)\,\delta^{(4)}(q - k_1 - k_2 - k_3) \\
&+ 3n_1(1+n_2)(1+n_3)\,\delta^{(4)}(q + k_1 - k_2 - k_3) \\
&+ 3n_1 n_2(1+n_3)\,\delta^{(4)}(q + k_1 + k_2 - k_3) \\
&+ n_1 n_2 n_3\,\delta^{(4)}(q + k_1 + k_2 + k_3)]\,.
\end{aligned} \tag{4.11}$$

Here

$$n_k = \frac{1}{e^{\omega_k/T} - 1} \tag{4.12}$$

is the paramagnon occupation number, and the four-dimensional δ-function describes energy and momentum conservation, $\delta^{(4)}(q + k_1 + k_2 + k_3) = \delta(\omega_q + \omega_1 + \omega_2 + \omega_3)\delta^{(3)}(\boldsymbol{q} + \boldsymbol{k}_1 + \boldsymbol{k}_2 + \boldsymbol{k}_3)$. The combinatorial factor $\mathbb{S} = 2(N + 2) = 10$ is due to summation over paramagnon polarisations.

4.2.2 The Diagrammatic Expansion

Here we wish to comment on the structure of diagrams included in the self-energy—the reader may safely skip this technical subsection. The loop diagrams in Fig. 4.4a contribute to the running coupling β_q. The external momentum is q, Λ_0 is the ultraviolet cutoff and the momentum in the loop is in the range $\Lambda_0 > p > q$. The diagrams in Fig. 4.4b contribute to the self-energy. All diagrams in Fig. 4.4 posses a quadratic, ultraviolet divergences. Such quadratic divergences lack physical meaning and are removed during the renormalisation scheme. Upon subtraction of the quadratically divergent contributions, the typical momentum in the *external loop* is $k \sim \Delta, T$ while the typical momentum in the *internal loop* is $\Lambda_0 > p > \Delta, T$. The internal loops of the double loop diagrams are shown inside dashed boxes in Fig. 4.4b(b and c). The series of internal loops in Fig. 4.4b coincides with the series of the running coupling constant shown in Fig. 4.4a. Hence we point out that the logarithmically divergent part of the *sunset diagram* Fig. 4.3, is already taken into account in our RG calculation of the mass gap Δ in Eq. (4.7). For example, Fig. 4.4b(c) is a part of the sunset diagram. In the diagrammatic series Fig. 4.4b we consider only the real part of the sunset diagram. We stress that the a central point of this chapter is the consideration of the imaginary part of the sunset diagram. And, to extract the most important physics relating to the imaginary part we will need to consider a different, infinite sub-series represented in Fig. 4.5. The following sections are dedicated to this point.

4.3 The Golden Rule of Quantum Kinetics

The discussion of the previous section regarding perturbation theory, RG and the decay width Eq. (4.11) relies on two conditions, (i) smallness of the coupling constant, $\beta \ll 1$, to justify a finite perturbative expansion, and (ii) a small paramagnon decay width to energy ratio $\Gamma \ll \Delta$, such that the notion of the thermal occupation number (4.12) is well defined. The thermal occupation number, Eq. (4.12), requires the quasi-particles to be on-mass-shell; $\omega = \omega_q = \sqrt{\boldsymbol{q}^2 + \Delta^2}$. However, broad quasiparticles do not satisfy this condition and instead their dispersion may (crudely speaking) fall

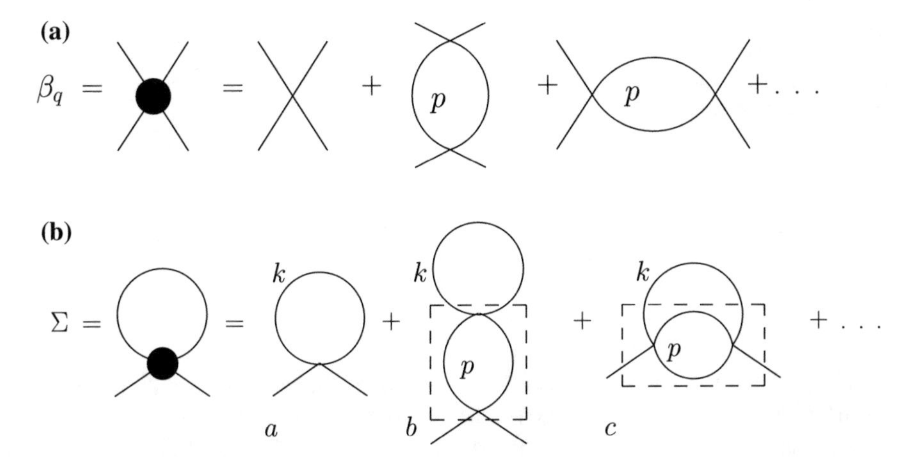

Fig. 4.4 Diagrammatic subseries: **a** Diagrams contributing to the coupling constant. **b** Diagrams contributing to the self-energy

anywhere in the range $\omega_{\boldsymbol{q}} - \Gamma/2 < \omega < \omega_{\boldsymbol{q}} + \Gamma/2$. For extremely broad quasiparticles $\Gamma/\Delta \sim 1$, the notion of a quasiparticle breaks down. Note for sufficiently large q such that $\Gamma_q(\omega = \omega_q) \ll \omega_q$, the quasiparticles are still well defined. Close to the Néel temperature where $T \gg \Delta \to 0$, point (ii) is not satisfied; the paramagnons experience overdamping due to the heat bath.

Having the above considerations in mind, we will now develop a technique appropriate to handle the regime of large heat bath scattering and the subsequent large deviation from equilibrium thermal occupation numbers Eq. (4.12)—we designate this the *hot quantum soup* regime. The hot quantum soup regime will certainly overlap with the crossover to the classical critical regime, however, we do persist with the terminology *classical critical*. Classical critical is appropriate to underline the dimensional crossover, 4D → 3D, and with it, the irrelevance of the time dimension. Chapter 5 explicitly considers this problem. Instead we use the term hot quantum soup to underline the broadening and overdamped dynamics of paramagnons in the presence of large heat bath occupation. Furthermore, it is not clear whether these two regimes exactly coincide.

The first step in our proposed technique is to dispense with the thermal occupation numbers and instead rewrite Eq. (4.11) in terms of spectral functions. In the limit of vanishing width, the imaginary part of the retarded Greens function is given simply as

$$-\frac{1}{\pi} Im\, G^R(\omega, \boldsymbol{q}) = \frac{1}{2\omega_q}\left[\delta(\omega - \omega_q) - \delta(\omega + \omega_q)\right] , \tag{4.13}$$

which follows from Eq. (4.4). Inserting this in Eq. (4.5) we find

$$S_{\boldsymbol{q}}(\omega) = \frac{1}{2\omega_q}\left[(1 + n_q)\delta(\omega - \omega_q) + n_q\delta(\omega + \omega_q)\right] . \tag{4.14}$$

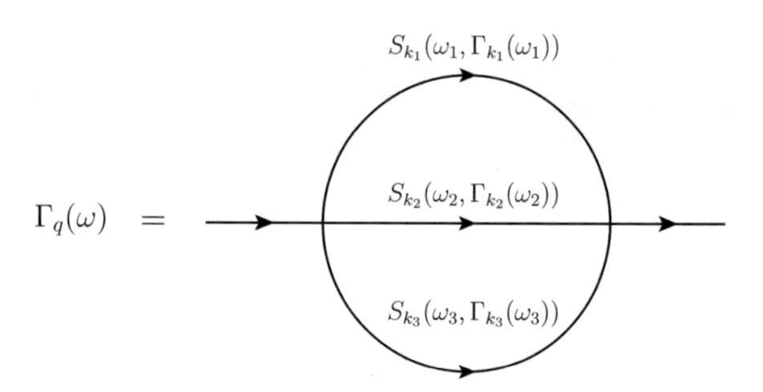

Fig. 4.5 Diagrammatic illustration of Dyson equation describing the golden rule of quantum kinetics

The first term in brackets in Eq. (4.14) describes the creation of a paramagnon from the heat bath, while the second term in describes the absorption of a paramagnon into the heat bath. We now use Eq. (4.14) to rewrite the width Eq. (4.11) as

$$\begin{aligned}\Gamma_q(\omega) &= \mathbb{S}(8\pi)^2\beta^2\frac{(1-e^{-\omega/T})}{2\omega}\int S_{k_1}(\omega_1)S_{k_2}(\omega_2)S_{k_3}(\omega_3)\\ &\times (2\pi)^4\delta(\omega-\omega_1-\omega_2-\omega_3)\delta(\boldsymbol{q}-\boldsymbol{k}_1-\boldsymbol{k}_2-\boldsymbol{k}_3)\\ &\times \frac{d\omega_1 d^3k_1}{(2\pi)^3}\frac{d\omega_2 d^3k_2}{(2\pi)^3}\frac{d\omega_3 d^3k_3}{(2\pi)^3}\,. \end{aligned} \tag{4.15}$$

The key improvement is that we have replaced occupation numbers (4.15), by the general expression (4.10) of the spectral function. The expression (4.15) is therefore valid for quasiparticles of significant broadness. Namely, it provides a valid description of the hot quantum soup regime where quasiparticles are poorly defined, $\Gamma \gtrsim \omega$. We name the closed set of equations, (4.10) and (4.15), the *golden rule of quantum kinetics*.

The self-consistent solution of the golden rule of quantum kinetics is a Dyson-equation-like procedure to determine the spectral function $S_{\boldsymbol{q}}(\omega,\Gamma_{\boldsymbol{q}})$. Figure 4.5 gives a diagrammatic representation of the Dyson equation. Most importantly, the spectral function can be directly compared with experiment. It is important to point out that the diagram shown in Fig. 4.5 is not a typical perturbative Feynman/Matsubara diagram; the lines in Fig. 4.5 are not Greens functions, they are spectral functions. In the summation of the golden rule of quantum kinetics, Fig. 4.5, it is the infrared divergences we seek to control. In contrast, and as discussed in Sect. 4.2.2, the summation of usual Feynman diagrams contributing to the mass and coupling constant renormalization was used to control the logarithmic ultraviolet divergences.

The self consistent solution of the golden rule of quantum kinetics provides the spectral and width functions, and is our primary result. We now list the key arguments and justifications of the method:

(i) We assume proximity, such that $\beta_q \ll 1$, to the quantum critical point to justify truncation of diagrams.

(ii) Upon approach to the Néel temperature, perturbation theory for the imaginary part suffers from infrared divergences. Hence failure of perturbation theory in this regime is not due to a large coupling constant, but instead due to the heat bath. As discussed, the golden rule of quantum kinetics removes the infrared, power-divergence in the overdamped regime.

(iii) Importantly, away from the Néel temperature the golden rule of quantum kinetics reduces to the Fermi golden rule.

(iv) The RG procedure accounts only for the on-mass-shell contribution to the real part of the sunset self-energy. However, in our evaluation of the imaginary part of the self-energy using Eqs. (4.15), (4.10), we consider both the on- and off-mass-shell contributions. One can exploit the analytic properties, i.e. via Kramers-Kronig relation, of our results to subsequently find the off mass-shell contribution to the real part of the self-energy. This extra step is beyond what is presented in the text, instead the calculation is performed in the Appendix C. We find that the off-mass-shell energy dependent contribution is negligibly small. Furthermore, away from the Néel temperature/overdamped regime, one does not need to consider the off mass-shell contribution at all.

We draw the readers attention to other approaches to the thermal field theory, see e.g. Refs. [7, 8]. These works do not rely on proximity to a QCP, and therefore the truncation of the Matsubara diagrams is uncontrolled.

4.4 Mathematical Analysis of the Golden Rule of Quantum Kinetics

We now undergo a mathematical analysis of the golden rule of quantum kinetics, to understand generic features without reference to any particular system. Our aim is to demonstrate the necessity of the golden rule of quantum kinetics (4.15) and (4.10). To facilitate the discussion and presentation, we disregard the RG running of the coupling constant and set it to a constant value

$$\beta = 0.2. \tag{4.16}$$

While in the next section we will explicitly reinstate the running.

We can perform some integrations within Eq. (4.15) analytically. Here we present equations only for $q = 0$ in order to avoid lengthy expressions,

$$\Gamma_{q=0}(\omega) = \frac{\mathbb{S}\beta^2}{\pi} \frac{\left(1 - e^{-\omega/T}\right)}{\omega} \int_{-\infty}^{+\infty} d\omega_1 d\omega_2 \int_0^{+\infty} dk_1^2 dk_2^2$$
$$\times \int_{(k_1-k_2)^2}^{(k_1+k_2)^2} dk_3^2 \, S_{k_1}(\omega_1) S_{k_2}(\omega_2) S_{k_3}(\omega - \omega_1 - \omega_2) \tag{4.17}$$

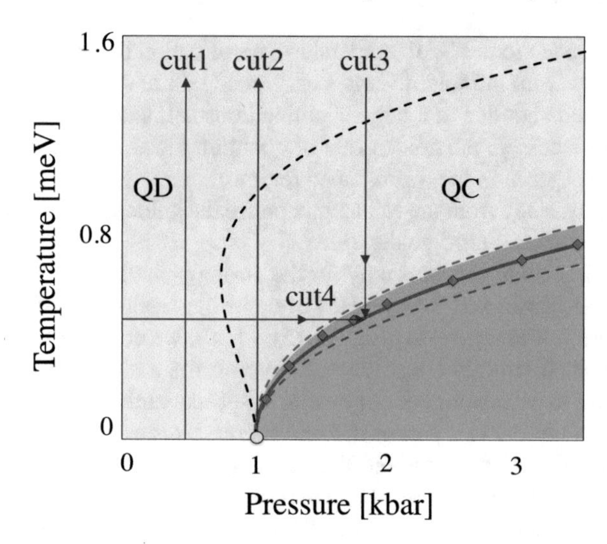

Fig. 4.6 The phase diagram with parameters for $TlCuCl_3$. The black dashed lines separate QD and QC regimes. The cuts: cut1, cut2, cut3, and cut4 represent traces through the phase diagram along which we evaluate the decay width

From here we rely on straightforward numerical evaluation of this expression.

First, we wish to evaluate along the trace, denoted *cut3* in Fig. 4.6, which approaches the Néel temperature from above at fixed pressure. Following this cut, we find it convenient to rescale by temperature, such that ω/T, Γ_q/T, Δ/T, and q/T are dimensionless. We also remind the reader that we have set the paramagnon speed to unity, $c = 1$, and hence momentum $q \to cq$ has the dimension of energy. Figure 4.7 plots of the paramagnon width function $\Gamma_{q=0}(\omega)$ versus ω for Δ/T ranging from $\Delta/T = 1$ to $\Delta/T = 0.1$.

Figure 4.7a, shows the width function $\Gamma_{q=0}(\omega)$ calculated using the Fermi golden rule (4.11). Figure 4.7b shows the width function calculated using the golden rule of quantum kinetics, i.e. by iterative solution of Eqs. (4.15) and (4.10). The two methods must collapse to the same result in the limit of vanishing Γ/ω. A comparison of Fig. 4.7 shows that they do indeed coincide in this limit. However, when considering small values of Δ and ω the two methods yield very different results. This is not surprising since the Fermi golden rule assumes the on-mass-shell notion related to Eq. (4.12). This notion, and also the Fermi golden rule fail at sufficiently small values of Δ/T where the width is very large, $\Gamma/\Delta > 1$. As we have stressed throughout this chapter, the golden rule of quantum kinetics does not require the on-mass-shell condition. For the remainder of our analysis we will use only the golden rule of quantum kinetics.

The spectral function $S_q(\omega)$, as given by Eq. (4.10), provides a direct access to experimental measurements. In Fig. 4.8a we present the spectral functions $S_{q=0}(\omega)$ which utilise the width functions $\Gamma_{q=0}(\omega)$ presented in Fig. 4.7. To balance the dimension of the spectral function, $[\text{Energy}]^{-2}$, we multiply by $S \to T^2 S$. To supplement

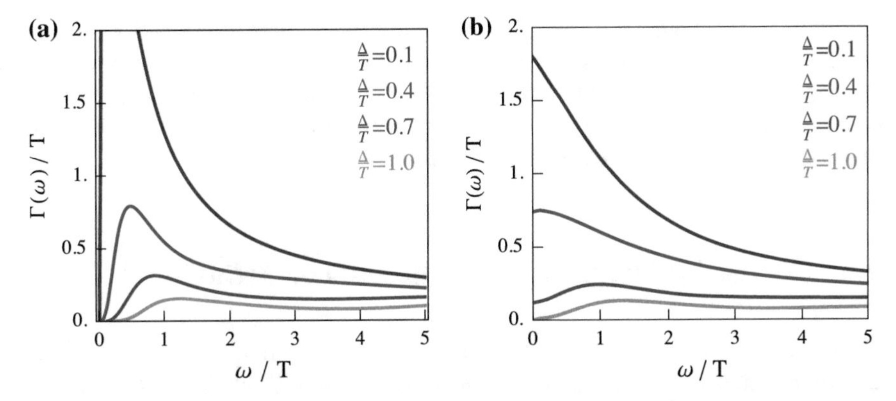

Fig. 4.7 Paramagnon width function at zero momentum, $\Gamma_{q=0}(\omega)$, versus frequency. The function is calculated with the coupling constant (4.16) for different values of the gap Δ. **a** Obtained using the simple Fermi golden rule, Eq. (4.11). **b** Obtained using the golden rule of quantum kinetics; Eqs. (4.15) and (4.10). Note the shifted origin on the ω/T-axis in (**a**)

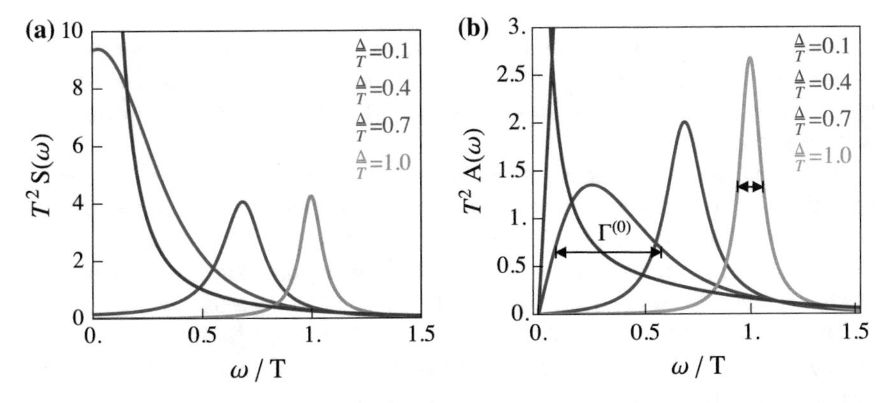

Fig. 4.8 **a** The spectral function $T^2 S_{q=0}(\omega)$ versus frequency for different values of the gap Δ. **b** The spectral density $A_{q=0}(\omega) = -\frac{1}{\pi} ImG^R(\omega, q = 0)$ versus frequency for different values of the gap Δ. Both $S_{q=0}(\omega)$ and $A_{q=0}(\omega)$ correspond to $\Gamma_{q=0}(\omega)$ (solid lines) in Fig. 4.7. Note the shifted origin on the ω/T-axis in (**b**)

the results shown in Fig. 4.8a, in Fig. 4.8b we present plots of the spectral density, $A_{\boldsymbol{q}}(\omega) = -\frac{1}{\pi} ImG^R(\omega, \boldsymbol{q})$. The spectral density $A(\omega)$ and spectral function share a simple relation, which is given in Eq. (4.5). It is the spectral density which been used to determine effective line widths from experimental data.

We define $\Gamma_q^{(0)}$ to be the full width at half maximum (FWHM) of the spectral density, which is represented by the arrowed lines in Fig. 4.8b. Note that this definition of $\Gamma_q^{(0)}$ has no ω-dependence, but it depends on the gap Δ, momentum $\boldsymbol{q}$, and temperature T.

We see from Fig. 4.8b that for sufficiently small values of Δ/T, the definition of $\Gamma^{(0)}$ as FWHM of the spectral density practically does not make sense—the spectral function $A(\omega)$ becomes asymmetric and even divergent. Such divergent behaviour indicates the crossover to the overdamped regime. The explicit crossover value of Δ_c depends on the value of the running coupling constant β_Λ. The smaller values of β_Λ correspond to the smaller Δ_c/T. The available experimental data for $TlCuCl_3$ lay within the regime of (reasonably) well defined $\Gamma^{(0)}$.

4.5 Comparison with Experimental Data on $TlCuCl_3$

We now compare the golden rule of quantum kinetics, in particular the width $\Gamma^{(0)}_{q=0}$, to the paramagnon widths obtained via neutron scattering in $TlCuCl_3$ [1]. To compare, we use the parameters derived for $TlCuCl_3$ in Chap. 2, which are

$$\begin{aligned} \beta_0 &= 0.23 \quad \text{for} \quad \Lambda_0 = 1\ \text{meV}\ , \\ p_c &= 1.01\,\text{kbar}\ , \quad \gamma = 0.68\,\text{meV/kbar}^{1/2}\ . \end{aligned} \tag{4.18}$$

Using parameters (4.18) and the theory developed in the present work we can calculate decay widths throughout the entire disordered phase diagram. In Chap. 2, the decay widths have been calculated along cut1 and cut2; they correspond to the perturbative regime.

In this chapter we consider cut3 and cut4 in Fig. 4.6. These cuts are coincident upon the Néel temperature and hence the *simple* perturbative RG used for cut1 and cut2 is not sufficient. We need RG plus the golden rule of quantum kinetics, Eqs. (4.15) and (4.10). In the vicinity of the Néel temperature spectral lines become asymmetric and hence the definition of width becomes ambiguous. We use values of $\Gamma^{(0)}$ defined in Sect. 4.4. In evaluating Eq. (4.15), the coupling β_Λ formally runs with energy scale $\Lambda = \max\{\sqrt{\omega^2 - q^2}, T\}$, yet we use $\Lambda = \max\{\Delta, T\}$, which makes a negligible difference [9].

In Fig. 4.9 we present theoretical and experimental values of the width $\Gamma^{(0)}_{q=0}$ and the gap Δ. Figure 4.9a corresponds to the vertical cut3 in Fig. 4.6, i.e. temperature varies at fixed pressure, $p = 1.75$ kbar. Figure 4.9b corresponds to the horizontal cut4 in Fig. 4.6, i.e. pressure varies at fixed temperature, $T = 0.5$ meV. Agreement between theoretical and experimental widths presented in Fig. 4.9a is very good. This includes the highly nontrivial, hot quantum soup regime close to the Néel temperature where the width calculated via the golden rule of quantum kinetics is different from that calculated via the simple Fermi golden rule. On the other hand, Fig. 4.9b demonstrates a disagreement between theory and experiment by a factor of two in the theoretically straightforward interval $0 < p < p_c$. In principle one can assign the disagreement to impurities. However, it is unlikely since the agreement seen in Chap. 2 for the endpoints of this interval, $p = 0$ in Fig. 2.9a and b and at $p = p_c$ in Fig. 2.9c and d is excellent. Hence, the reason for the disagreement remains unclear to the author.

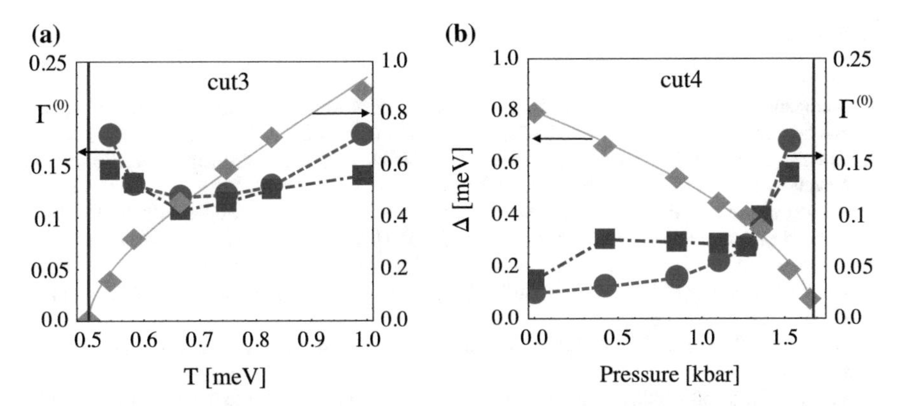

Fig. 4.9 Theoretical and experimental values of the width $\Gamma^{(0)}_{q=0}$ and the gap Δ. Panel a corresponds to the vertical cut3 in Fig. 4.6, temperature varies at fixed pressure, $p = 1.75$ kbar. Panel b corresponds to the horizontal cut4 in Fig. 4.6, pressure varies at fixed temperature, $T = 0.5$ meV. In both panels blue circles show theoretical results of the present work while magenta squares show experimental results of Ref. [1]. Yellow diamonds show experimental results for the gap [1]. Dashed blue and magenta as well as solid yellow lines connecting the points are given just for guidance

4.6 Conclusions

Physically, the problem we consider was stimulated by the observation of magnetic excitations (paramagnons) in the magnetically disordered phase of the three dimensional, dimerised quantum antiferromagnet $TlCuCl_3$ [1]. The data in Ref. [1] is taken over a range of pressures and temperatures that go beyond the regime of validity of standard perturbation theory, i.e. close to the Néel temperature where the paramagnon width becomes comparable to its energy due to multiple scattering events from other paramagnons in the heat bath.

Motivated by this particular issue, we formulate a generic nonperturbative technique that cures the enigmas of perturbative approaches, and offers reliable results in the physically interesting regimes of hot, dense mediums. The developed technique provides an economical approach to this class of problems and possesses two key properties: (i) it regulates the infrared behaviour via a resummation of medium effects i.e. the self-consistent inclusion of line-shapes, and (ii) allows one to handle the calculation of non-equilibrium responses at finite temperature.

We perform an explicit comparison with data on $TlCuCl_3$, which yields excellent agreement between the theory and experiment. We stress that the formulation is generic and applicable to any quantum field theory with weak coupling.

References

1. Merchant P, Normand B, Kramer KW, Boehm M, McMorrow DF, Rüegg C (2014) Quantum and classical criticality in a dimerized quantum antiferromagnet. Nat Phys 10(5):373–379
2. Qin YQ, Normand B, Sandvik AW, Meng YZ (2015) Multiplicative logarithmic corrections to quantum criticality in three-dimensional dimerized antiferromagnets. Phys Rev B 92:214401
3. Rüegg C, Furrer A, Sheptyakov D, Strässle T, Krämer KW, Güdel H-U, Mélési L (2004) Pressure-induced quantum phase transition in the spin-liquid $TlCuCl_3$. Phys Rev Lett 93:257201
4. Sachdev S (2011) Quantum phase transitions. Cambridge University Press
5. Affleck I, Wellman GF (1992) Longitudinal modes in quasi-one-dimensional antiferromagnets. Phys Rev B 46:8934–8953
6. Scammell HD, Sushkov OP (2015) Asymptotic freedom in quantum magnets. Phys Rev B 92:220401
7. van Hees H, Knoll J (2001) Renormalization in self-consistent approximation schemes at finite temperature: theory. Phys Rev D 65:025010
8. van Hees H, Knoll J (2002) Renormalization of self-consistent approximation schemes at finite temperature. II. Applications to the sunset diagram. Phys Rev D 65:105005
9. For off-mass shell four momentum $\mu^2 = \omega^2 - q^2 \neq \Delta^2$, the only significant contribution to Eq. (4.15) comes from the 'window' $\mu^2 \approx \Delta^2 \pm \Gamma^2$, since integrand (4.15) is heavily suppressed otherwise. In the limit $\Delta \ll T, \Gamma > \Delta$ but the running scale will be set by $\Lambda = \max\{\mu, T\} = T$. In the opposite limit $\Delta \gg T$, then $\Gamma \ll \Delta$ since $\Gamma \sim e^{-\Delta/T}$ (see Sect. 4) and the running scale is essentially unaffected; $\Lambda^2 \approx \Delta^2 \pm \Gamma^2 \approx \Delta^2$

Chapter 5
Dimensional Reduction in Quantum Critical Systems

Abstract This chapter discusses dimensional reduction in 3+1D antiferromagnets. The observable of interest is the order parameter (staggered magnetisation). At zero temperature, the order parameter is well described by a 3+1D theory. In the near vicinity of the Néel temperature (T_N), the system is expected to crossover into a regime whereby the order parameter is described by a theory in one lower dimension. This is 'dimensional reduction'. A recent analysis of the order paramter versus temperature, performed on the data for 3+1D antiferromagnet $TlCuCl_3$ [1], has identified the crossover boundaries to the dimensionally reduced regime. It was found that on approach to T_N a relatively sharp crossover occurs, whereby the order parameter near T_N displays scaling in accord with the 3D theory. In this chapter we analyse the same scenario. Starting from a $T = 0$, 3+1D quantum field theory, and perturbatively including temperature, we quantitatively describe the experimental data on $TlCuCl_3$ across the range $0 \leq T \leq T_N$. Our theory cannot reproduce the 3D scaling exponents in the limit $T \to T_N$, yet it fully accounts for the observed sharp crossover behaviour in $TlCuCl_3$. Ultimately, we expect observables to scale with exponents of a lower dimensional theory, however, our results provide a new perspective on the enigmatic problem of dimensional reduction.

5.1 Introduction

Dimensional reduction, as defined in the present context, is a phenomenon attributed to the interplay of quantum and thermal fluctuations in the vicinity of quantum phase transitions, see e.g. [2, 3]. Dimensional reduction, also referred to as the quantum-to-classical crossover, belongs to a class of finite-temperature crossovers, which also includes the quantum-critical crossover discussed in Chap. 4. Finite-temperature crossovers in the vicinity of quantum phase transitions are a rich subject alluring a great deal of attention [3–33], although many aspects either remain poorly understood or have lacked quantitative tests against experimental or numerical data.

There is an important technical complication with dimensional reduction, that is, standard field theoretic RG approaches fail since the presence of non-zero temperature introduces infrared divergences into certain classes of Feynman diagrams as the

H. Scammell, *Interplay of Quantum and Statistical Fluctuations in Critical Quantum Matter*, Springer Theses,
https://doi.org/10.1007/978-3-319-97532-0_5

temperature approaches T_N. For a general mathematical discussion of this important issue see Chap. 36 of Ref. [34], or for a discussion directly relevant to this chapter see Appendix D.

We are solely concerned with the phenomenon of dimensional reduction in 3+1 dimensional quantum critical systems; i.e. $d : 3 + 1 \rightarrow 3$, and explicitly the realisation in 3+1 dimensional quantum antiferromagnets. It has long been argued that 3+1D antiferromagnets (i.e. 3D *quantum* antiferromagnets) undergo a dimensional reduction or quantum-to-classical crossover in the vicinity of the Néel temperature transition line, see Fig. 5.1.

The key idea being that for parametrically large temperatures, $T \gg \Delta$, thermal fluctuations are occurring at a characteristic length scale, $\beta = 1/T$, that is much smaller than the characteristic time scales of quantum fluctuations, $\xi_\tau \sim 1/\Delta$. Note, here the dynamic critical exponent is $z = 1$. In this regime the system behaves as though it is uncorrelated along the time dimension and, accordingly, may be considered as a classical statistical ensemble in one lower dimension.

Alternatively, starting directly from a 4D (Euclidean) statistical theory, finite temperature can be introduced as a truncation along one of the four spatial axes, as $\beta = 1/T$. Similar arguments can be made about characteristic length scales such that for $T \gg \Delta$, the system acts as a 3D statistical theory. These statements about crossovers occurring in 3+1D and 4D are essentially equivalent when discussing static observables. Since this chapter solely considers static observables, we henceforth discuss both 3+1D and 4D systems on an equal footing.

Physically, if an effective dimensional reduction were to take place, then observables, such as order parameter, gaps, etc. would show scaling with critical indices

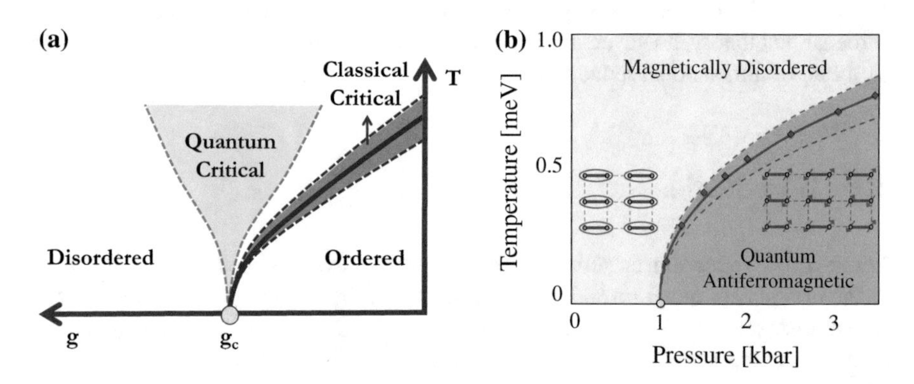

Fig. 5.1 Phase diagram and finite-temperature crossovers in 3+1 dimensional quantum critical systems. **a** Generic phase diagram with quantum and thermal tuning axes given by g and T, respectively. Two finite-temperature crossover regions are indicated; the quantum critical regime and the classical critical regime. Dimensional reduction corresponds to the crossover into the classical critical regime. **b** Experimental realisation of a 3+1 dimensional quantum critical system; dimerised quantum antiferromagnet $TlCuCl_3$. Quantum tuning parameter corresponds to hydrostatic pressure, p. Red shaded region indicates the classical critical regime. Solid red line schematically represents the Néel temperature, while data for Néel temperature are taken from [35]

of the lower dimensional critical theory. Experimental data for the 3+1 dimensional antiferromagnet $TlCuCl_3$ [36] provides the order parameter as a function of temperature, and runs from very low temperatures $T \approx 0$, up to $T \approx T_N$, and hence, at least in principle, contains information on the dimensional reduction. Based on fitting asymptotic scaling forms to this data, it has been argued recently in Ref. [1], that in the vicinity of T_N the data fits the scaling predictions of a 3D critical theory. From this analysis, lines of dimensional reduction for the pressure-temperature phase diagram of $TlCuCl_3$ have been deduced. An illustration of the deduced dimensional reduction crossover lines is provided in Fig. 5.1a and b by the dashing red curves. While this fitting procedure supports the scenario of dimensional reduction, it provides no analytic insight into the behaviour of order parameter throughout the temperature range $0 \leq T \leq T_N$. In particular, it is not clear how observables scale in the intermediate regime between the, 3+1D at $T = 0$ and 3D at $T = T_N$ limits.

The present work provides a quantitative description of the order parameter over the entire range, $0 \leq T \lesssim T_N$, for the general $O(N)$, 3+1 dimensional critical theory, and explicitly applies the results to $TlCuCl_3$.

5.2 Asymptotic Scaling Forms

Let us briefly recall the RG results for the scaling and critical exponents of the order parameter, φ_c, for the generic $O(N)$ Euclidean field theory in $d = 4 - \epsilon$ dimensions. To first-order in the ϵ-expansion one obtains [34],

$$\varphi_c \sim (1-t)^{\beta_d}, \qquad \beta_d = \frac{1 - \frac{1}{2}\epsilon}{2 - \frac{N+2}{N+8}\epsilon}, \tag{5.1}$$

where $t = T/T_N$, such that $(1 - t)$ is a measure of detuning from the $4 - \epsilon$ dimensional classical critical point T_N. Hence, for the $O(3)$ model, in three and four dimensions one obtains $\beta_3 = 1/(2(2 - 5/11))$, and $\beta_4 = 1/2$, respectively.

The problem of dimensional reduction does not simply amount to different scaling regimes; $\varphi_c \sim (1-t)^{\beta_4}$ at $t \to 0$, to $\varphi_c \sim (1-t)^{\beta_3}$ at $t \to 1$. It is expected that under dimensional reduction, scaling near the thermal critical point is given by $\varphi_c \sim (1-t)^{\beta_3}$ at $t \to 1$. However, at $t \to 0$ one cannot expect the scaling $\varphi_c \sim (1-t)^{\beta_4}$, since this is now *far* from the thermal critical point and so scaling behaviour is not governed by this fixed point. Instead, at $t \to 0$ one returns to the 3+1D quantum critical theory whereby critical scaling is measured in terms of detuning, g, from the quantum critical point, g_c. To make this more explicit, the $t \to 0$, 3+1D critical scaling of the order parameter in detuning, g, is given by,

$$\varphi_c \sim (1 - g/g_c)^{\beta_4}, \tag{5.2}$$

$$\varphi_c \sim (1 - g/g_c)^{\beta_4} \left|\ln\left((1 - g/g_c)/b_1\right)\right|^{\frac{3}{N+8}}. \tag{5.3}$$

The second line takes into account the multiplicative logarithmic corrections expected at the upper critical dimension, and b_1 is a constant completely determined by the parameters of the field theory. The point being that there are two relevant perturbations, corresponding to the axes of the phase diagram Fig. 5.1; quantum detuning, g/g_c, and thermal detuning, t, and that it is poorly understood how $\varphi_c(g, T)$ scales for a given $g < g_c$ under thermal detuning throughout the range $t \in (0, 1)$. To this end, one may hypothesise a double scaling ansatz of the form,

$$\varphi_c \sim (1 - g/g_c)^{\beta_4}(1 - t)^{\beta_3} , \tag{5.4}$$

which would account for both the 3+1D quantum and 3D thermal scaling limits. But how to obtain such scaling behaviour from field theory?

The theoretical purpose of this chapter is to derive the analytic form of $\varphi_c(g, T)$ appropriate to describe the entire ordered phase, with range $g < g_c$ and $t \in (0, 1)$. We will show in the next section that it takes the form

$$\varphi_c(g, t) \sim (1 - g/g_c)^{\beta_4} \left[1 - t^2 f(g, t)\right]^{\beta_4} , \tag{5.5}$$

$$\varphi_c(g, t) \sim (1 - g/g_c)^{\beta_4} \left[1 - t^2 f(g, t)\right]^{\beta_4} \left|\ln\left((1 - g/g_c))\left[1 - t^2 f(g, t)\right]/b_2\right)\right|^{\frac{3}{N+8}} , \tag{5.6}$$

here $t = t(g)$, since T_N is dependent on detuning g, as per Chaps. 2 and 3. The function $f(g, t)$ plays a crucial role in describing the crossover physics going from $t \to 0$ to $t \to 1$. Explicit form of $f(g, t)$ is given in Sect. 5.3. Again, the second line takes into account multiplicative logarithmic corrections.

The scaling form of $\varphi_c(g, t)$ in Eq. (5.9) has the expected critical exponent for the quantum detuning, g/g_c, in 3+1D, i.e. $(1 - g/g_c)^{\beta_4}$. On the other hand, for thermal detuning, t, the critical exponent does not replicate the expected value for a 3D theory. Instead, thermal detuning maintains the 3+1D critical exponent, i.e. β_4 in $\left[1 - t^2 f(g, t)\right]^{\beta_4}$. However, the function $f(g, t)$ displays a non-analytic cusp at $t \to 1$, see Fig. 5.2, and as a result has a dramatic influence on the scaling of $\varphi_c(g, t)$ in the regime of $t \to 1$. The next section explicitly evaluates $\varphi_c(g, t)$ and $f(g, t)$ for parameters specific to 3+1D antiferromagnet $TlCuCl_3$.

5.3 Results and Analysis

We now have in mind comparing to experimental data on $TlCuCl_3$, as such the quantum tuning parameter is given by the applied pressure $p = -g$. Following from Chap. 2, the order parameter is given by [37],

$$\varphi_c^2(p, T) = \frac{1}{\alpha_0}\gamma^2(p - p_c)\left[\frac{\alpha_\Lambda}{\alpha_0}\right]^{\frac{-6}{N+8}} - (N-1)\sum_{\mathbf{k}} \frac{1/k}{(e^{\frac{k}{T}} - 1)} - 3\sum_{\mathbf{k}} \frac{1/\omega_{\mathbf{k}}}{(e^{\frac{\omega_{\mathbf{k}}}{T}} - 1)} , \tag{5.7}$$

where $\omega_{\boldsymbol{k}} = \sqrt{\boldsymbol{k}^2 + \Delta_H^2}$, and the (Higgs) gap is given by $\Delta_H^2 = 2\alpha_\Lambda \varphi_c^2$. From here onwards we specialise to $N = 3$. Note, we have excluded the thermal damping ansatz for this analysis, i.e. $\Gamma \to 0$, which appeared in Chap. 2. This choice allows for a simpler presentation, and does not significantly affect our conclusions.

It is instructive to present the Néel temperature,

$$T_N(p)^2 = \frac{12\gamma^2(p - p_c)}{5\alpha_0} \left[\frac{\alpha_0}{\alpha_{\Lambda(p,T_N)}}\right]^{\frac{6}{11}} , \tag{5.8}$$

which is utilised to rewrite Eq. (5.7) into a form suggestive of the critical scaling behaviour,

$$\varphi_c^2(t, p) = \frac{1}{\alpha_0}\gamma^2(p - p_c)\left[\frac{\alpha_\Lambda}{\alpha_0}\right]^{\frac{-6}{11}} \left[1 - t^2 f(t, p)\right] , \tag{5.9}$$

$$f(t, p) = \frac{12}{5}\left(2\sum_{\mathbf{x}} \frac{1/x}{(e^x - 1)} + 3\sum_{\mathbf{x}} \frac{1/y_{\boldsymbol{x}}}{(e^{y_{\boldsymbol{x}}} - 1)}\right)\left[\frac{\alpha_{\Lambda(p,t=1)}}{\alpha_{\Lambda(p,t)}}\right]^{\frac{-6}{11}} . \tag{5.10}$$

Here $\boldsymbol{x} = \boldsymbol{k}/T$, $y_{\boldsymbol{x}} = \omega_{\boldsymbol{k}}/T$, and now the running scale is written as a function of p and t, $\Lambda = \Lambda(p, t)$. Although in this chapter we alter the running scale to be,

$$\Lambda(p, t) = \sqrt{t^2 T_N^2 + \frac{1}{2}\Delta_H(p, tT_N)^2} , \tag{5.11}$$

instead of $\Lambda(p, t) = \max[tT_N, \frac{1}{\sqrt{2}}\Delta_H(p, tT_N)]$, as has been used previously—Chaps. 2, 3 and 4. This subtle change avoids unphysical cusps in the running coupling constant α_Λ, as seen for example in Figure 2.8. At the same time, this form reproduces the same results as Chaps. 2 and 3 in the important limits; $T = 0$, $\Lambda = 1/\sqrt{2}\Delta_H(p, 0)$ and at $T = T_N$, $\Lambda = T_N(p)$. Values of the fitting parameters were found in Chap. 2,

$$p_c = 1.01 \text{ kbar}, \ \gamma = 0.68 \text{ meV/kbar}^{1/2}, \ \frac{\alpha_0}{8\pi} = 0.23 . \tag{5.12}$$

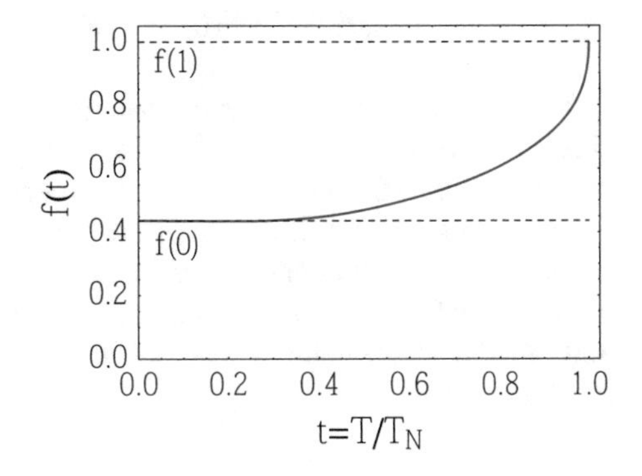

Fig. 5.2 The scaling function $f(t) = f(p, t)$ with $p = 2.5$ kbar from Eq. (5.10), as a function of thermal detuning t. $f(t)$ displays a non-analytic cusp at $t = 1$. The limits $f(0)$ and $f(1)$ are indicated by dashed lines

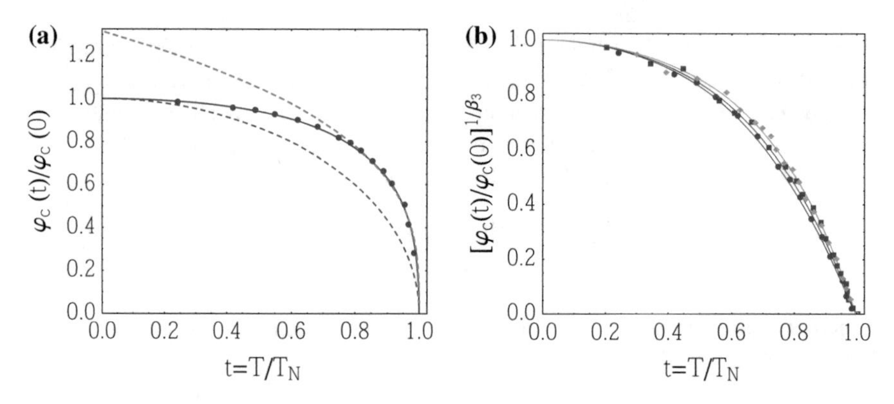

Fig. 5.3 Normalised order parameter. **a** Blue line shows the analytic results derived in this chapter (5.13), the red dashed line shows the asymptotic 4D scaling form (5.14), and the turquoise dashed line shows the asymptotic 3D scaling form (5.15). Normalisation is chosen such that the 4D asymptote (5.14) is fitted to the $t \to 0$ limit, while 3D asymptote (5.15) is fit to the $t \to 1$ limit, the different limits correspond to where asymptotes are expected to hold. Experimental data taken from $TlCuCl_3$ is shown by blue markers [36]. The data is normalised to unity in the $t = 0$ limit. **b** Comparison of analytic results (5.13) with experimental data [36], for various pressures $p = \{2.5, 3.5, 4.79\}$ kbar, shown by blue, maroon, and yellow markers, respectively. The results (5.13) and data [36] are raised to the power of $1/\beta_3$ to demonstrate the approximately linear scaling at $t \to 1$

In Fig. 5.3a we compare the normalised order parameter to the expected 3D scaling as well as experimental data [36]. The data, taken at various fixed pressures $p = \{2.5, 3.5, 4.79\}$ kbar, provides the order parameter as a function of temperature from $T \sim 0$ to $T \sim T_N$ and hence contains the desired information on the scaling with respect to thermal detuning. To compare the thermal scaling behaviour and to give an understanding to the role played by $f(p,t)$ as well as logarithmic corrections, we plot the following three curves,

$$\varphi_c(t,p)/\varphi_c(0,p) = \left[1 - t^2 f(t,p)\right]^{1/2} \left[\frac{\alpha_{\Lambda(p,t)}}{\alpha_{\Lambda(p,0)}}\right]^{\frac{-3}{11}}, \tag{5.13}$$

$$\varphi_c(t,p)/\varphi_c(0,p) = \left[1 - t^2\right]^{1/2}, \tag{5.14}$$

$$\mathcal{N}_0 \varphi_c^{3D}(t)/\varphi_c^{3D}(0) = \mathcal{N}_0 (1-t)^{\beta_3}. \tag{5.15}$$

Here (5.13) represents our full solution, (5.14) sets $f(p,t) = 1$ and omits logarithmic corrections, and (5.15) is the expected 3D scaling limit. Normalisation factor $\mathcal{N}_0$ is chosen to fit the $t \to 1$ limit of experimental data [36], where 3D scaling is expected to hold. All three curves are plotted in Fig. 5.3a, with solid blue, dashed maroon, and dashed turquoise lines, respectively for (5.13), (5.14), and (5.15). Experimental data [36] are given by blue markers. The difference between dashed maroon and solid blue lines demonstrates the combined significance of $f(p,t)$ and logarithmic corrections.

Agreement between our analytic result (5.13) and experimental data [36], blue line and points in Fig. 5.3, respectively, is our key result. We see that our analytic, 3+1D theory of the order parameter tracks the sharp crossover displayed in the data. Moreover, agreement between scaling forms of blue (5.13) and turquoise (5.15) lines over the range, $0.8 \leq t \leq 1$, is a remarkable result. The agreement demonstrates an *apparent* change of critical exponent, i.e. $\beta_4 \rightarrow \beta_3$, induced by $f(p, t)$ and the multiplicative logarithmic corrections.

We remind the reader that ultimately we expect the critical exponent of the order parameter to obey 3D predictions, and hence our results would give the wrong critical exponent asymptotically close to $t \rightarrow 1$. Instead, the primary result of this chapter is having an analytic description of $\varphi_c(p, t)$ appropriate to describe $0 \leq t \lesssim 1$. To this end, agreement between theory and experiment in Fig. 5.3a and b is very convincing. Of course, it remains the task of future high precision experimental or numerical studies to provide tests of our quantitative predictions for the entire (p, t) parameter range. And in particular, how close to the $t = 1$ limit does our theory still remain a good description?

A final, minor observation follows from (5.9)—it has been cast into a seemingly pressure independent, universal form, yet due to the p-dependence of $f(p, t)$ and within the logarithmic corrections, explicit dependence on pressure p can not be eliminated. We therefore expect the rescaled theoretical and experimental plots for $\varphi_c(p, t)/\varphi_c(p, 0)$ evaluated at various pressures to fail to exhibit a universal collapse. To demonstrate this point, Fig. 5.3b plots the rescaled $[\varphi_c(p, t)/\varphi_c(p, 0)]^{1/\beta_3}$ as a function of thermal tuning t. Raised to the power of $1/\beta_3$ one expects approximately linear scaling at $t \rightarrow 1$. We plot for three different values of p, and indeed see a small but noticeable departure from a universal collapse. The values of pressure are chosen to match the experimental values [36], and are $p = \{2.5, 3.5, 4.79\}$ kbar. Once again, agreement between theory and experiment is very compelling.

5.4 Concluding Remarks and Future Research

The novelty of the present work is that we have provided a continuous, analytic description of the order parameter across the entire range $0 \leq T \leq T_N$. Most importantly, our work has demonstrated that the sharp crossover-like behaviour observed for the order parameter is completely described by our 3+1D theory. Ultimately, we expect our results to give the wrong critical exponent in the limit $T \rightarrow T_N$. Nonetheless, our results provide a new perspective on the enigmatic problem of dimensional reduction. We therefore hope to inspire future unbiased QMC studies at non-temperatures to directly test our quantitive predictions over the entire interval, $0 \leq T \leq T_N$.

References

1. Merchant P, Normand B, Kramer KW, Boehm M, McMorrow DF, Rüegg C (2014) Quantum and classical criticality in a dimerized quantum antiferromagnet. Nat Phys 10(5):373–379
2. Sachdev S (2011) Quantum phase transitions. Cambridge University Press
3. Sachdev S (1997) Theory of finite-temperature crossovers near quantum critical points close to, or above, their upper-critical dimension. Phys Rev B 55:142–163
4. Scammell HD, Sushkov OP (2017) Nonequilibrium quantum mechanics: a "hot quantum soup" of paramagnons. Phys Rev B 95:024420
5. Singh KK (1975) Renormalization-group approach to a Bose system. Phys Rev B 12:2819–2823
6. Hertz JA (1976) Quantum critical phenomena. Phys Rev B 14:1165–1184
7. Suzuki M (1976) Relationship between d-dimensional quantal spin systems and (d + 1)-dimensional Ising systems equivalence, critical exponents and systematic approximants of the partition function and spin correlations. Prog Theor Phys 56(5):1454
8. Theumann WK (1980) Gaussian-to-Heisenberg crossover behavior of the specific heat in renormalized perturbation theory. Phys Rev B 21:1930–1940
9. Elderfield DJ (1980) Parametric representation of the dilute polymer system to O(ϵ^2): crossover from Wilson-Fisher to Gaussian fixed point as the Flory temperature is approached. J Phys C Solid State Phys 13(32):5883
10. Nicoll JF, Bhattacharjee JK (1981) Crossover functions by renormalization-group matching: O(ϵ^2) results. Phys Rev B 23:389–401
11. Creswick RJ, Wiegel FW (1983) Renormalization theory of the interacting Bose fluid. Phys Rev A 28:1579–1586
12. Guo H, Jasnow D (1987) Hyperuniversality and the renormalization group for finite systems. Phys Rev B 35:1846–1850
13. Eisenriegler E, Tomaschitz R (1987) Helmholtz free energy of finite spin systems near criticality. Phys Rev B 35:4876–4887
14. Rasolt M, Stephen MJ, Fisher ME, Weichman PB (1984) Critical behavior of a dilute interacting Bose fluid. Phys Rev Lett 53:798–801
15. Fisher DS, Hohenberg PC (1988) Dilute Bose gas in two dimensions. Phys Rev B 37:4936–4943
16. Chakravarty S, Halperin BI, Nelson DR (1989) Two-dimensional quantum Heisenberg antiferromagnet at low temperatures. Phys Rev B 39:2344–2371
17. Millis AJ (1993) Effect of a nonzero temperature on quantum critical points in itinerant fermion systems. Phys Rev B 48:7183–7196
18. Sokol A, Pines D (1993) Toward a unified magnetic phase diagram of the cuprate superconductors. Phys Rev Lett 71:2813–2816
19. Sachdev T, Senthil T, Shankar R (1994) Finite-temperature properties of quantum antiferromagnets in a uniform magnetic field in one and two dimensions. Phys. Rev. B 50:258–272
20. Zülicke U, Millis AJ (1995) Specific heat of a three-dimensional metal near a zero-temperature magnetic phase transition with dynamic exponent z=2, 3, or 4. Phys Rev B 51:8996–9004
21. Sachdev S, Chubukov AV, Sokol A (1995) Crossover and scaling in a nearly antiferromagnetic fermi liquid in two dimensions. Phys Rev B 51:14874–14891
22. Ioffe LB, Millis AJ (1995) Critical behavior of the uniform susceptibility of a fermi liquid near an antiferromagnetic transition with dynamic exponent z = 2. Phys Rev B 51:16151–16158
23. Sengupta AM, Georges A (1995) Non-fermi-liquid behavior near a T = 0 spin-glass transition. Phys Rev B 52:10295–10302
24. Sachdev S, Read N, Oppermann R (1995) Quantum field theory of metallic spin glasses. Phys Rev B 52:10286–10294
25. Chubukov AV, Pines D, Stojkovic BP (1996) Temperature crossovers in cuprates. J Phys Condens Matter 8(48):10017
26. Liao S-B, Strickland M (1997) Dimensional crossover and effective exponents. Nucl Phys B 497(3):611–638

27. Lawrie ID (1993) Critical phenomena in field theories at finite temperature. J Phys A Math Gen 26(23):6825
28. O'Connor D, Stephens CR (1991) Phase transitions and dimensional reduction. Nucl Phys B 360(2):297–336
29. O'Connor D, Stephens CR (1994) Crossover scaling: a renormalization group approach. Proc R Soc Lond A Math Phys Eng Sci 444(1921):287–296
30. Freire F, O'Connor D, Stephens CR (1994) Dimensional crossover and finite-size scaling below T_c. J Stat Phys 74(1):219–238
31. O'Connor D, Stephens CR (1994) Effective critical exponents for dimensional crossover and quantum systems from an environmentally friendly renormalization group. Phys Rev Lett 72:506–509
32. O'Connor D, Stephens CR (1994) Environmentally friendly renormalization. Int J Mod Phys A 09(16):2805–2902
33. O'Connor D, Stephens CR (2002). Renormalization group theory of crossovers. Phys Rep 363(46):425–545
34. Zinn-Justin J (2002) Quantum field theory and critical phenomena. International series of monographs on physics. Clarendon Press
35. Rüegg C, Furrer A, Sheptyakov D, Strässle T, Krämer KW, Güdel H-U, Mélési L (2004) Pressure-induced quantum phase transition in the spin-liquid $TlCuCl_3$. Phys Rev Lett 93:257201
36. Rüegg C, Normand B, Matsumoto M, Furrer A, McMorrow DF, Krämer KW, Güdel HU, Gvasaliya SN, Mutka H, Boehm M (2008) Quantum magnets under pressure: controlling elementary excitations in $TlCuCl_3$. Phys Rev Lett 100:205701
37. Scammell HD, Sushkov OP (2015) Asymptotic freedom in quantum magnets. Phys Rev B 92:220401

Chapter 6
Continuity of the Order Parameter in Magnetic Condensates

Abstract Phase transitions captured by the spontaneous symmetry breaking (SSB) mechanism are necessarily second-order. Magnetic field induced phase transitions in quantum antiferromagnets are associated with the SSB-condensation of bosonic triplet excitations and are hence expected to be continuous. However, theoretical descriptions of the non-zero temperature, magnetic field induced condensation of triplons have erroneously predicted a discontinuity in the order parameter. This theoretical issue dates back $\sim$50 years to the work of Fadeev and Popov on the dilute Bose gas condensation, and ultimately arises due to the perturbative treatment not respecting symmetries of the action. The present work approaches the problem starting from a relativistic quantum field theory and demonstrates how perturbative corrections are to be handled in a method consistent with the underlying symmetries. A key result is that our treatment satisfies the Goldstone theorem, which is shown to be a necessary condition for the continuity of the phase transition.

6.1 Introduction

The order parameter, arising in symmetry broken phases, is an indispensable concept in the study of critical phenomena. For broken continuous symmetries, along with the order parameter, there must exist gapless modes, called Goldstone excitations, which reflect the original symmetry of the system. This is a statement of the Goldstone theorem [1]. For the system under consideration in this chapter—three dimensional, $O(3)$ quantum antiferromagnets—there exists three tuning handles for the SSB, the quantum tuning parameter g (explicitly we consider pressure, $p = -g$), the magnetic field B, and the temperature, T. Condensation under the separate tuning of $p > p_c$ or $B > B_c$ corresponds to the spontaneous breakdown of either $O(2)$ or $O(3)$, and supports either one or two Goldstone modes, respectively. In three dimensions, the ordered phases driven by p and/or B survive up to a non-zero Néel temperature T_c. Here we use T_c instead of T_N, for generality.

In real compounds, i.e. $TlCuCl_3$, $CsFeCl_3$, and others, tuning magnetic field and/or pressure offers two unique scenarios to study SSB, the order parameter, and Goldstone physics. This has by now attracted immense experimental [2–9] and theo-

H. Scammell, *Interplay of Quantum and Statistical Fluctuations in Critical Quantum Matter*, Springer Theses,
https://doi.org/10.1007/978-3-319-97532-0_6

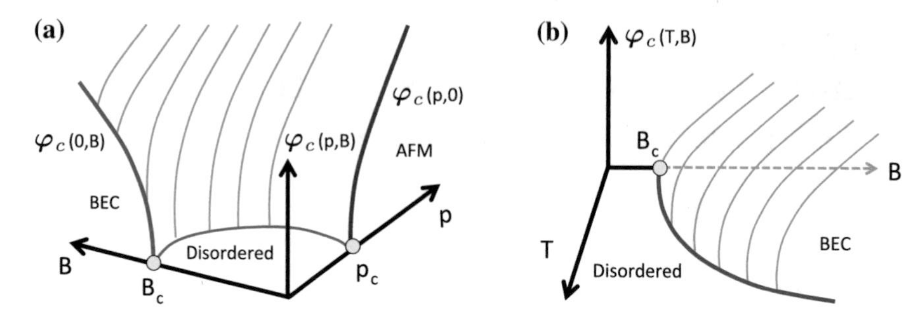

Fig. 6.1 Schematic of the *continuous* order parameter, $\varphi_c(p, B, T)$, across the critical (p, B, T) phase diagram. **a** The zero-temperature order parameter, $\varphi_c(p, B)$. The red line emanating from the QCP, $(p_c, 0)$, designates the pure AFM order parameter, as discussed in previous chapters. The blue lines show the evolution of the order parameter under non-zero magnetic field, and are hence in the BEC universality. **b** The non-zero temperature order parameter, $\varphi_c(T, B)$. This figure corresponds to $p < p_c$, since $B_c(T = 0 > 0)$, the yellow point. At larger temperatures, larger magnetic field is required to establish the condensate

retical [10–14] interest over the past two decades. Despite the long history, there remains a fundamental issue within theoretical approaches to the finite temperature, field induced phase transition, i.e. the $(p < p_c, B, T)$ sector. Previous theoretical approaches, which were concentrated at the $(p < p_c, B, T)$ sector, have employed the dilute Bose gas model [11, 15] and/or bond-operator technique [16]. In the end, these techniques rely on the Hartree-Fock- Popov approximation developed ~50 years ago in the context of the dilute Bose gas [17, 18]. However, it is known that the Hartree-Fock-Popov approximation breaks down in the vicinity of the critical temperature T_c and subsequently predicts a first-order phase transition [19]. For the dilute gas problem this is manifest in the discontinuity of the density, while for the magnetic BEC, there is a discontinuity in the staggered magnetisation [11].

The statement of interest of the present work is to construct a theory that naturally allows for the continuity of the order parameter in magnetic condensates across the entire critical (p, B, T) phase diagram, as depicted in Fig. 6.1. This equivalently allows for a truly second-order phase transition. Results will be explicitly discussed in terms of magnetic systems, yet the technique/method employed is of generic importance for systems described by $O(N)$ models, with $N > 1$. This chapter also serves as a theoretical basis for future chapters, Chap. 7 explores the interesting aspects of the extended phase diagram (p, B, T), while Chap. 8 looks at the decay channels of the Higgs mode in the ordered phases of the (p, B, T) diagram.

The quantum phase transition between ordered and disordered phases is described by the effective field theory with the following Lagrangian [20–22],

$$\mathscr{L} = \frac{1}{2}(\partial_t\vec{\varphi} - \vec{\varphi} \times \vec{B})^2 - \frac{1}{2}(\vec{\nabla}\vec{\varphi})^2 - \frac{1}{2}m_0^2\vec{\varphi}^2 - \frac{1}{4}\alpha_0\vec{\varphi}^4. \tag{6.1}$$

The vector field $\vec{\varphi}$ describes staggered magnetisation, B is an external applied field, and we set $g\mu_B = c = 1$. We now briefly outline the *tree-level* phase transitions captured by this Lagrangian. Consider first $B = 0$, the pressure induced QPT results from tuning the mass term, m_0^2, for which we take the linear expansion $m_0^2(p) = \gamma^2(p_c - p)$, where $\gamma^2 > 0$ is a coefficient and p is the applied pressure. Varying the pressure leads to two distinct phases; (i) for $p < p_c$ we have $m_0^2 > 0$, and the classical expectation value of the field is zero $\varphi_c^2 = 0$. This describes the magnetically disordered phase, the system has a global rotational symmetry, and the excitations are gapped and triply degenerate. (ii) For pressures $p > p_c$ we have $m_0^2 < 0$, and the field obtains a non-zero classical expectation value $\varphi_c^2 = |m_0^2|/\alpha_0$. This describes the magnetically ordered, antiferromagnetic phase. Varying m_0^2 from positive to negative spontaneously breaks the $O(3)$ symmetry of the system. In the broken phase there are two gapless Goldstone excitations, and one gapped Higgs excitation.

Next consider non-zero B at fixed $p < p_c$: For $B < m$ the system has $O(2)$ symmetry, and the degeneracy of the triplet modes is lifted by Zeeman splitting. The field induced QPT results from tuning $B > m_0$. The order parameter field is $\varphi_c^2 = \left(B^2 - m_0^2\right)/\alpha_0$, and there exists a single gapless Goldstone mode.

6.2 Disordered Phase

We now discuss how to go beyond mean-field. Everywhere in the text $m_0^2 = \gamma_0^2(p_c - p)$ and α_0 represent the zero temperature mass tuning parameter and coupling constant without quantum fluctuation corrections. Taking into account quantum and thermal fluctuation corrections due to interaction term $\frac{1}{4}\alpha_0\vec{\varphi}^4$, we will denote the renormalised parameters $m_0^2 \to m_{\Lambda,\sigma}^2$ and $\alpha_0 \to \alpha_\Lambda$. The explicit form for $m_{\Lambda,\sigma}^2 = m_{\Lambda,\sigma}^2(p, T, B)$ depends on the location within the phase diagram, and polarisation $\sigma = +1, 0, -1$. Explicit expressions will be derived shortly and are presented in Eqs. (6.4) and (6.5).

In the disordered phase the Euler-Lagrange equation with (6.1) results in the following dispersion

$$\omega_k^\sigma = \sqrt{k^2 + m_{\Lambda,\sigma}^2} + \sigma B. \tag{6.2}$$

Importantly, the σB term is not renormalised. This is a consequence of a Ward identity (Larmor theorem). While the Lagrangian can be diagonalised by fields that represent the physical states with corresponding dispersion (6.2), we find it convenient to calculate fluctuation corrections in the Cartesian basis $\vec{\varphi} = (\varphi_x, \varphi_y, \varphi_z)$. Let us denote by $\mathcal{V}$ the part of the Lagrangian (6.1) independent of derivatives. Then, using a Wick decoupling of the interaction term $\frac{1}{4}\alpha_0\vec{\varphi}^4$, in the single-loop approximation we find

$$\frac{\partial^2 \mathscr{V}}{\partial \varphi_x^2} = m_0^2 - B^2 + 3\alpha_0 \langle \varphi_x^2 \rangle + \alpha_0 \langle \varphi_y^2 \rangle + \alpha_0 \langle \varphi_z^2 \rangle$$
$$\frac{\partial^2 \mathscr{V}}{\partial \varphi_y^2} = m_0^2 - B^2 + \alpha_0 \langle \varphi_x^2 \rangle + 3\alpha_0 \langle \varphi_y^2 \rangle + \alpha_0 \langle \varphi_z^2 \rangle$$
$$\frac{\partial^2 \mathscr{V}}{\partial \varphi_z^2} = m_0^2 + \alpha_0 \langle \varphi_x^2 \rangle + \alpha_0 \langle \varphi_y^2 \rangle + 3\alpha_0 \langle \varphi_z^2 \rangle \tag{6.3}$$

where $\langle \varphi_x^2 \rangle$ is the loop integral over the Green's function of field φ_x. The loops encode both quantum and thermal corrections. The Greens functions and explicit expressions for the loops will be presented in the Appendix E. An explicit calculation shows $\langle \varphi_x^2 \rangle = \langle \varphi_y^2 \rangle$, hence from Eq. (6.3) we have rather trivially satisfied the $O(2)$ Ward identity: $\partial^2 \mathscr{V}/\partial \varphi_x^2 - \partial^2 \mathscr{V}/\partial \varphi_y^2 = 0$. Throughout this chapter we use the following *loop*-notation to denote the renormalised mass parameters,

$$m_{\Lambda,\pm}^2 \equiv m_0^2 + 3\alpha_0 \langle \varphi_x^2 \rangle + \alpha_0 \langle \varphi_y^2 \rangle + \alpha_0 \langle \varphi_z^2 \rangle \,, \tag{6.4}$$
$$m_{\Lambda,0}^2 \equiv m_0^2 + \alpha_0 \langle \varphi_x^2 \rangle + \alpha_0 \langle \varphi_y^2 \rangle + 3\alpha_0 \langle \varphi_z^2 \rangle \,, \tag{6.5}$$

as opposed to carrying out the RG procedure. This choice is for notational convenience—it allows for transparent algebraic manipulations. Of course we have in mind that at the final step of evaluation, RG will be applied. Exactly what is meant by this statement is made clear in the Appendix E, where the expressions corresponding to Eqs. (6.4) and (6.5), after the application of RG, are provided.

Beginning from the disordered phase, we can derive the critical magnetic field, including temperature dependence, by evaluating the gapless point in the dispersion of the critical mode (6.2). Explicitly setting the critical mode of Eq. (6.2) to zero, i.e. $\omega_{\mathbf{0}}^-(B_c^-) = 0$, and substituting Eq. (6.4), one finds

$$B_c^-(T) = \sqrt{m_0^2 + \alpha_0 (4\langle \varphi_x^2 \rangle + \langle \varphi_z^2 \rangle)} \,. \tag{6.6}$$

We have taken advantage of the $O(2)$ symmetry property $\langle \varphi_x^2 \rangle = \langle \varphi_y^2 \rangle$, and once again note that RG must be applied to the expression (6.6). The superscript, '−', in B_c^- indicates the approach to the critical field from the disordered phase. Next, B_c^+ will indicate the approach from the ordered phase.

6.3 Ordered BEC Phase

Consider next the BEC phase, $B > B_c(T)$. The vector field is then written $\vec{\varphi} = (\varphi_c + \sigma, \pi, z)$, where φ_c is the order parameter field, and the Lagrange-fields σ and π correspond to hybridisations of the true Higgs and Goldstone modes. Further implications of this hybridisation will be discussed in Chap. 8. The Lagrange-field,

z, directly corresponds to the precession mode. The physical (diagonal) modes of the system have dispersion,

$$\omega_{\boldsymbol{k}}^{H} = \sqrt{\boldsymbol{k}^2 + 3B^2 - m_{\Lambda,H}^2 + \sqrt{4B^2\boldsymbol{k}^2 + (3B^2 - m_{\Lambda,H}^2)^2}}\,, \tag{6.7}$$

$$\omega_{\boldsymbol{k}}^{G} = \sqrt{\boldsymbol{k}^2 + 3B^2 - m_{\Lambda,H}^2 - \sqrt{4B^2\boldsymbol{k}^2 + (3B^2 - m_{\Lambda,H}^2)^2}}\,, \tag{6.8}$$

$$\omega_{\boldsymbol{k}}^{z} = \sqrt{\boldsymbol{k}^2 + m_{\Lambda,z}^2}\,. \tag{6.9}$$

The superscripts $\{H, G, z\}$ designate the Higgs, Goldstone, and precession modes, respectively. The renormalised mass terms of the ordered phase; $m_0^2 \to m_{\Lambda,H}^2$ and $m_0^2 \to m_{\Lambda,z}^2$, are analogous to those of the disordered phase Eqs. (6.4) and (6.5), and are defined below in Eqs. (6.12) and (6.14).

We write an effective potential, $\mathscr{V}$, from the non-derivative terms of the Lagrangian (6.1) including first-order in α_0 loop corrections, and expand about the minimum φ_c,

$$\begin{aligned}\mathscr{V} = &-\frac{1}{2}\left(B^2 - m_0^2\right)(\varphi_c + \sigma, \pi, z)^2 + \frac{1}{2}\sigma^2\left\{3\alpha_0\langle\sigma^2\rangle + \alpha_0\langle\pi^2\rangle + \alpha_0\langle z^2\rangle\right.\\ &\left.+ 18\alpha_0^2\varphi_c^2\langle\sigma^2\sigma^2\rangle + 2\alpha_0^2\varphi_c^2\langle\pi^2\pi^2\rangle + 2\alpha_0^2\varphi_c^2\langle z^2z^2\rangle\right\} + \frac{1}{2}\pi^2\left\{\alpha_0\langle\sigma^2\rangle + 3\alpha_0\langle\pi^2\rangle\right.\\ &\left.+ \alpha_0\langle z^2\rangle + 4\alpha_0^2\varphi_c^2\langle\sigma^2\pi^2\rangle\right\} + \frac{1}{2}z^2\left\{\alpha_0\langle\sigma^2\rangle + \alpha_0\langle\pi^2\rangle + 3\alpha_0\langle z^2\rangle + 4\alpha_0^2\varphi_c^2\langle\sigma^2z^2\rangle\right\}\\ &+ \frac{1}{2}\varphi_c^2\left\{3\alpha_0\langle\sigma^2\rangle + \alpha_0\langle\pi^2\rangle + \alpha_0\langle z^2\rangle\right\} + \frac{1}{4}\varphi_c^4\,.\end{aligned} \tag{6.10}$$

The single bracket notation $\langle\pi^2\rangle$ denotes the loop integral over internal π-field propagator. Loop integrals and propagators are explicitly evaluated in the Appendix E. Similarly, the trace over an internal line is given by $\langle\sigma^2\rangle$ and $\langle z^2\rangle$ for the σ-field and z-component loops, respectively. In the effective potential, we also need to include terms/traces $\langle\sigma^2\pi^2\rangle$, $\langle\sigma^2z^2\rangle$, $\langle\pi^2\pi^2\rangle$, and $\langle z^2z^2\rangle$. Such contributions are rendered first-order in α_0 due to the presence of the order parameter field squared, which is of the order $\mathscr{O}(\alpha_0^{-1})$ i.e. $\alpha_0^2\varphi_c^2 \to \alpha_0\left(B^2 - m_0^2\right)$. Numerical pre-factors in (6.10) arise from permutation symmetry of the loops, for a general discussion see e.g. [23].

Minimisation of the effective potential $\partial\mathscr{V}/\partial\varphi_c = 0$, defines the renormalised order parameter,

$$\varphi_c^2 = \frac{B^2 - m_0^2}{\alpha_0} - 3\langle\sigma^2\rangle - \langle\pi^2\rangle - \langle z^2\rangle. \tag{6.11}$$

This is an important result, and shows how the meanfield order parameter, $\varphi_c^2 = \left(B^2 - m_0^2\right)/\alpha_0$, is renormalised to leading-order in α_0. Having established the renormalised order parameter, we can now evaluate the curvature of the effective potential (6.10), with respect to the field *directions*, σ, π and z,

$$
\begin{aligned}
\frac{\partial^2 \mathscr{V}}{\partial \sigma^2} &= 3\alpha_0\varphi_c^2 - (B^2 - m^2) + 3\alpha_0\langle\sigma^2\rangle + \alpha_0\langle\pi^2\rangle + \alpha_0\langle z^2\rangle + 18\alpha_0^2\varphi_c^2\langle\sigma^2\sigma^2\rangle \\
&\quad + 2\alpha_0^2\varphi_c^2\langle\pi^2\pi^2\rangle + 2\alpha_0^2\varphi_c^2\langle z^2 z^2\rangle \\
&= 2\left(B^2 - m_0^2 - \alpha_0\left\{3\langle\sigma^2\rangle + \langle\pi^2\rangle + \langle z^2\rangle\right\}\right) + 2\alpha_0^2\varphi_c^2\left\{9\langle\sigma^2\sigma^2\rangle + \langle\pi^2\pi^2\rangle + \langle z^2 z^2\rangle\right\} \\
&\equiv 2(B^2 - m_{\Lambda,H}^2)\,, \qquad (6.12)\\
\frac{\partial^2 \mathscr{V}}{\partial \pi^2} &= (m^2 - B^2) + \alpha_0\varphi_c^2 + \alpha_0\langle\sigma^2\rangle + 3\alpha_0\langle\pi^2\rangle + \alpha_0\langle z^2\rangle + 4\alpha_0^2\varphi_c^2\langle\sigma^2\pi^2\rangle \\
&= 2\alpha_0\langle\pi^2\rangle - 2\alpha_0\langle\sigma^2\rangle + 4\alpha_0^2\varphi_c^2\langle\sigma^2\pi^2\rangle\,, \qquad (6.13)\\
\frac{\partial^2 \mathscr{V}}{\partial z^2} &= m^2 + \alpha_0\varphi_c^2 + \alpha_0\langle\sigma^2\rangle + \alpha_0\langle\pi^2\rangle + 3\alpha_0\langle z^2\rangle + 4\alpha_0^2\varphi_c^2\langle\sigma^2 z^2\rangle \\
&= B^2 + 2\alpha_0\langle z^2\rangle - 2\alpha_0\langle\sigma^2\rangle + 4\alpha_0^2\varphi_c^2\langle\sigma^2 z^2\rangle \\
&\equiv m_{\Lambda,z}^2\,. \qquad (6.14)
\end{aligned}
$$

Simplifications are obtained by the substitution of the renormalised order parameter field (6.11). Finally, we obtain the desired expressions for the renormalised mass terms in the ordered phase,

$$
\begin{aligned}
m_{\Lambda,H}^2 &= m_0^2 + 3\alpha_0\langle\sigma^2\rangle + \alpha_0\langle\pi^2\rangle + \alpha_0\langle z^2\rangle - \alpha_0^2\varphi_c^2\left\{9\langle\sigma^2\sigma^2\rangle + \langle\pi^2\pi^2\rangle + \langle z^2 z^2\rangle\right\}, \qquad (6.15)\\
m_{\Lambda,z}^2 &= B^2 + 2\alpha_0\langle z^2\rangle - 2\alpha_0\langle\sigma^2\rangle + 4\alpha_0^2\varphi_c^2\langle\sigma^2 z^2\rangle. \qquad (6.16)
\end{aligned}
$$

Again, these quantities appear in the dispersion relations for modes of the ordered phase Eqs. (6.7), (6.8) and (6.9).

6.4 Results

We are now in a position to demonstrate the continuity of the order parameter as well as verify the Goldstone theorem.

6.4.1 *Order Parameter*

First, it is necessary that all modes continuously evolve into their counterpart at the phase transition $B = B_c^{\pm}(T)$,

$$
\text{(I)}: \omega_k^+ = \omega_k^H\,, \quad \text{(II)}: \omega_k^0 = \omega_k^z, \quad \text{(III)}: \omega_k^- = \omega_k^G\,. \qquad (6.17)
$$

It is a straightforward task to check the continuity of the modes, i.e. to confirm (6.17), and we leave it to Appendix E to do so explicitly.

Performing an RG treatment of (6.11), one obtains

$$\varphi_c^2(T) = \frac{1}{\alpha_\Lambda}(B^2 - m_0^2\left[\frac{\alpha_\Lambda}{\alpha_0}\right]^{\frac{5}{11}}) - \sum_{\mathbf{k}}\left\{\frac{2(5B^2 - m_{\Lambda,H}^2)}{(\omega_{\mathbf{k}}^H)^2 - (\omega_{\mathbf{k}}^G)^2}\left[\frac{n(\omega_{\mathbf{k}}^H)}{\omega_{\mathbf{k}}^H} - \frac{n(\omega_{\mathbf{k}}^G)}{\omega_{\mathbf{k}}^G}\right]\right.$$
$$\left. + 2\left[\frac{n(\omega_{\mathbf{k}}^H)}{\omega_{\mathbf{k}}^H} + \frac{n(\omega_{\mathbf{k}}^G)}{\omega_{\mathbf{k}}^G}\right] + \frac{n(\omega_{\mathbf{k}}^z)}{\omega_{\mathbf{k}}^z}\right\}, \tag{6.18}$$

where $n(\omega_{\mathbf{k}}) = 1/(e^{\frac{\omega_{\mathbf{k}}}{T}} - 1)$. The order parameter (6.18) is an analytic function of its arguments, (p, B, T). To demonstrate the continuity of the order parameter across the non-zero temperature phase transition, it suffices to show that there is a unique critical field $B_c^-(T) = B_c^+(T)$ such that critical mode of the disordered phase, and the order parameter both identically vanish, i.e.

$$\omega_{\mathbf{0}}^-(B_c^-) = \varphi_c(B_c^+) = 0\,. \tag{6.19}$$

To demonstrate, we evaluate the critical magnetic field by finding the root of Eq. (6.11),

$$B_c^+(T) = \sqrt{m_0^2 + \alpha_0(3\langle\sigma^2\rangle + \langle\pi^2\rangle + \langle z^2\rangle)}\,. \tag{6.20}$$

whereas attacking from the disordered phase we obtained Eq. (6.6). At $B = B_c^\pm$, the loop integrals of the z-field become identical $\langle\varphi_z^2\rangle = \langle z^2\rangle$, which can be simply understood since the dispersions and Greens functions identically match. Hence, to demonstrate $B_c^-(T) = B_c^+(T)$, it suffices to show that,

$$3\langle\sigma^2\rangle + \langle\pi^2\rangle = 4\langle\varphi_x^2\rangle\,. \tag{6.21}$$

Evaluating the loop integrals at B_c, we confirm the condition (6.21),

$$\begin{aligned}
3\langle\sigma^2\rangle + \langle\pi^2\rangle &= \int\frac{d^3k}{(2\pi)^3}\left\{\frac{(10B^2 - 2m_{\Lambda,H}^2)}{\omega_H^2 - \omega_G^2}\left[\frac{n(\omega_H)}{\omega_H} - \frac{n(\omega_G)}{\omega_G}\right] + 2\left[\frac{n(\omega_H)}{\omega_H} + \frac{n(\omega_G)}{\omega_G}\right]\right\} \\
&= 2\int\frac{d^3k}{(2\pi)^3}\frac{1}{\sqrt{k^2 + B^2}}\left\{n(\omega_{\mathbf{k}}^H) + n(\omega_{\mathbf{k}}^G)\right\}\,, \quad (\text{at } B = B_c\,) \\
&= 2\int\frac{d^3k}{(2\pi)^3}\frac{1}{\omega_{\mathbf{k}}^0}\left\{n(\omega_{\mathbf{k}}^+) + n(\omega_{\mathbf{k}}^-)\right\} \\
&= 4\langle\varphi_x^2\rangle\,.
\end{aligned}$$

To obtain the penultimate line, we use the continuity conditions of the mode dispersions at $B_c^\pm$, Eq. (6.17). We have therefore successfully obtained a second-order phase transition at finite temperature. This is a key result of the chapter.

6.4.2 Goldstone Theorem

Another crucial result is that the present technique satisfies the Goldstone theorem. At meanfield level the physical Goldstone mode is gapless. A proper treatment of loop quantum and thermal corrections leaves such a mode gapless. The effective potential (6.10) contains the first-order in α_0 corrections, but is written in the non-physical (i.e. non-diagonal) Cartesian basis. In this basis, to respect the Goldstone theorem it is necessary and sufficient to show that the π-field represents the flat direction in the field space, i.e. $\partial^2\mathscr{V}/\partial\pi^2 = 0$, Eq. (6.13). Loop integrals are printed below, from which it is an algebraic task to verify that anywhere in the ordered phase,

$$\frac{\partial^2\mathscr{V}}{\partial\pi^2} = 2\alpha\langle\pi^2\rangle - 2\alpha\langle\sigma^2\rangle + 4\alpha^2\varphi_c^2\langle\sigma^2\pi^2\rangle = 0. \tag{6.22}$$

The first two loop integrals are found, in Appendix E, to be

$$\langle\sigma^2\rangle = \int\frac{d^3k}{(2\pi)^3}\left\{\frac{\left[6B^2 - 2m_{\Lambda,H}^2 + \omega_H^2 - \omega_G^2\right]}{\omega_H^2 - \omega_G^2}\frac{n_H}{2\omega_H} - \frac{\left[6B^2 - 2m_{\Lambda,H}^2 - \omega_H^2 + \omega_G^2\right]}{\omega_H^2 - \omega_G^2}\frac{n_G}{2\omega_G}\right\},$$

$$\langle\pi^2\rangle = \int\frac{d^3k}{(2\pi)^3}\left\{\frac{\left[2B^2 + 2m_{\Lambda,H}^2 + \omega_H^2 - \omega_G^2\right]}{\omega_H^2 - \omega_G^2}\frac{n_H}{2\omega_H} - \frac{\left[2B^2 + 2m_{\Lambda,H}^2 - \omega_H^2 + \omega_G^2\right]}{\omega_H^2 - \omega_G^2}\frac{n_G}{2\omega_G}\right\}. \tag{6.23}$$

The final loop integral in Eq. (6.22) is found to be,

$$\begin{aligned}\langle\sigma^2\pi^2\rangle = \mathrm{Re}\int&\frac{d^3k}{(2\pi)^3}\frac{1}{2\omega_H 2\omega_G}\\ &\times\left\{\left[(1+n_H)(1+n_G) - n_H n_G\right]\left[\frac{1}{i0 - \omega_H - \omega_G} - \frac{1}{i0 + \omega_H + \omega_G}\right]\right.\\ &\left.+\left[(1+n_H)n_G - n_H(1+n_G)\right]\left[\frac{1}{i0 - \omega_H + \omega_G} - \frac{1}{i0 + \omega_H - \omega_G}\right]\right\}.\end{aligned} \tag{6.24}$$

Note, we are dealing with the real part of the loop diagram since we are concerned with mass corrections. Great care must be taken when obtaining this expression, since one has to account for contributions of the anomalous Greens functions. A complete discussion of Greens functions and loop integrals is provided in Appendix E.

Adding the loop integrals (6.23) and (6.24), and performing some straightforward but lengthy algebra, one confirms (6.22). Once it is verified that the curvature of the effective potential remains flat under perturbations, it immediately follows that the physical, i.e. diagonal, Goldstone mode of the system has a gapless dispersion Eq. (6.8).

6.5 Discussion and Conclusion

In general, it is a theoretical challenge to quantitatively describe the influence of non-zero temperatures on a quantum system. This is certainly the case when considering the non-zero temperature, magnetic field induced Bose-condensation in three dimensional quantum antiferromagnets. Perturbative approaches based on the Hartree-Fock-Popov technique fail to describe the order parameter in the vicinity of the non-zero temperature transition. The present work instead starts with a relativistic quantum field theory, and demonstrates a general scheme to handle quantum and thermal perturbations to generate a consistent description of the order parameter and of the excitations. The essential result is that under renormalisation, we maintain the Goldstone theorem, i.e. the Goldstone excitations remain gapless. Ultimately, we find this to be the necessary ingredient to obtain continuity of the order parameter. Having established a consistent theoretical description of the order parameter throughout the (p, B, T) phase diagram, this chapter forms a basis for future explorations of magnon Bose-condensates. Chapters 7 and 8 are dedicated to this exploration.

References

1. Goldstone J, Salam A, Weinberg S (1962) Broken symmetries. Phys Rev 127:965–970
2. Kurita N, Tanaka H (2016) Magnetic-field- and pressure-induced quantum phase transition in $CsFeCl_3$ proved via magnetization measurements. Phys Rev B 94:104409
3. Ch Ruegg N, Cavadini A, Furrer H-U, Gudel K, Kramer H, Mutka A, Wildes KH, Vorderwisch P (2003) Bose-Einstein condensation of the triplet states in the magnetic insulator $TlCuCl_3$. Nature 423(6935):62–65
4. Jaime M, Correa VF, Harrison N, Batista CD, Kawashima N, Kazuma Y, Jorge GA, Stern R, Heinmaa I, Zvyagin SA, Sasago Y, Uchinokura K (2004) Magnetic-field-induced condensation of triplons in Han purple pigment $BaCuSi_2O_6$. Phys Rev Lett 93:087203
5. Kofu M, Kim J-H, Ji S, Lee S-H, Ueda H, Qiu Y, Kang H-J, Green MA, Ueda Y (2009) Weakly coupled S = 1/2 quantum spin singlets in $Ba_3Cr_2O_8$. Phys Rev Lett 102:037206
6. Wang Z, Quintero-Castro DL, Zherlitsyn S, Yasin S, Skourski Y, Islam ATMN, Lake B, Deisenhofer J, Loidl A (2016) Field-induced magnonic liquid in the 3D spin-dimerized antiferromagnet $Sr_3Cr_2O_8$. Phys Rev Lett 116:147201
7. Okada M, Tanaka H, Kurita N, Johmoto K, Uekusa H, Miyake A, Tokunaga M, Nishimoto S, Nakamura M, Jaime M, Radtke G, Saúl A (2016) Quasi-two-dimensional Bose-Einstein condensation of spin triplets in the dimerized quantum magnet $Ba_2CuSi_2O_6Cl_2$. Phys Rev B 94:094421
8. Grundmann H, Sabitova A, Schilling A, von Rohr F, Förster T, Peters L (2016) Tuning the critical magnetic field of the triplon Bose-Einstein condensation in $Ba_{3-x}Sr_xCr_2O_8$. New J Phys 18(3):033001
9. Brambleby J, Goddard PA, Singleton J, Jaime M, Lancaster T, Huang L, Wosnitza J, Topping CV, Carreiro KE, Tran HE, Manson ZE, Manson JL (2017) Adiabatic physics of an exchange-coupled spin-dimer system: magnetocaloric effect, zero-point fluctuations, and possible two-dimensional universal behavior. Phys Rev B 95:024404
10. Giamarchi T, Tsvelik AM (1999) Coupled ladders in a magnetic field. Phys Rev B 59:11398–11407
11. Nikuni T, Oshikawa M, Oosawa A, Tanaka H (2000) Bose-Einstein condensation of dilute magnons in $TlCuCl_3$. Phys Rev Lett 84:5868–5871

12. Matsumoto M, Normand B, Rice TM, Sigrist M (2002) Magnon dispersion in the field-induced magnetically ordered phase of $TlCuCl_3$. Phys Rev Lett 89:077203
13. Matsumoto M, Normand B, Rice TM, Sigrist M (2004) Field- and pressure-induced magnetic quantum phase transitions in $TlCuCl_3$. Phys. Rev. B 69:054423
14. Utesov OI, Syromyatnikov AV (2014) Theory of field-induced quantum phase transition in spin dimer system $Ba_3Cr_2O_8$. J Magn Magn Mater 358–359:177–182
15. Kawashima N (2005) Critical properties of condensation of field-induced triplet quasiparticles. J Phys Soc Jpn 74(Suppl):145–150
16. Sirker J, Weiße A, Sushkov OP (2005) The field-induced magnetic ordering transition in $TlCuCl_3$. J Phys Soc Jpn 74(Suppl):129–134
17. Popov VN (1965) JETP 20:1185
18. Popov VN, Fadeev LD (1965) JETP 20:890
19. Shi H, Griffin A (1998) Finite-temperature excitations in a dilute Bose-condensed gas. Phys Rep 304(1–2):1–87
20. Fisher DS (1989) Universality, low-temperature properties, and finite-size scaling in quantum antiferromagnets. Phys Rev B 39:11783–11792
21. Sachdev S (2011) Quantum phase transitions. Cambridge University Press
22. Affleck I (1991) Bose condensation in quasi-one-dimensional antiferromagnets in strong fields. Phys Rev B 43:3215–3222
23. Kleinert H, Schulte-Frohlinde V (2001) Critical properties of ϕ^4-theories. World Scientific

Chapter 7
Multiple Universalities in Order-Disorder Magnetic Phase Transitions

Abstract We consider isotropic quantum antiferromagnets under applied magnetic field, pressure and temperature. The combination of these three tuning handles leads to an extended phase diagram which has not been explored within a single approach. Employing a quantum field theoretic approach that allows us to consider the entire extended phase diagram, we predict the emergence of multiple (three) universalities under combined pressure and field tuning. Changes of universality are signalled by changes of the critical exponent ϕ. Explicitly, we predict the existence of two new exponents $\phi = 1$ and 1/2 as well as recovering the known exponent $\phi = 3/2$. We also predict logarithmic corrections to the power law scaling.

7.1 Introduction

In the study of critical phenomena, a central goal is to uncover and categorise the universal features. Understanding the universal features allows for a more powerful and enlightening description of the complex system at hand. A key property of systems in the vicinity of a critical point is the corresponding critical exponents, which govern the scaling behaviour of observables. A great effort—experimental, numerical and theoretical—has been devoted to uncovering new universal behaviour and critical exponents.

The present work considers three dimensional (3D) quantum antiferromagnets (QAF), where the combined interplay between pressure, magnetic field and temperature (p, B, T) provides a unique and experimentally achievable arena to explore new universal behaviour. We are concerned with the critical exponent ϕ which governs the critical field-critical temperature power law,

$$\text{(a)}: \ \delta B_{\text{BEC}} \sim T^{\phi}, \qquad \text{(b)}: \ \delta T_N \sim B^{1/\phi}, \tag{7.1}$$

The shift of the BEC transition line at small temperature is shown schematically in Fig. 7.1a, while the shift of the AFM/Néel transition line at small field is shown in Fig. 7.1b. We also provide Fig. 7.2 which depicts the three dimensional (p, B, T) phase diagram.

H. Scammell, *Interplay of Quantum and Statistical Fluctuations in Critical Quantum Matter*, Springer Theses,
https://doi.org/10.1007/978-3-319-97532-0_7

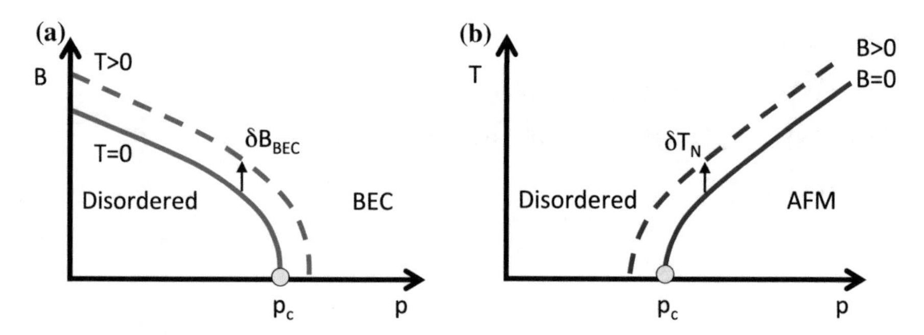

Fig. 7.1 Critical field and temperature power law shifts. **a** Shift of critical field-pressure line with temperature $\delta B_{\mathrm{BEC}} \sim T^{\phi}$. Solid blue curve is at zero temperature, dashed blue at non-zero temperature. **b** Shift of critical (Néel) temperature-pressure line with field $\delta T_N \sim B^{1/\phi}$. Solid red curve is at zero field, dashed red at non-zero field

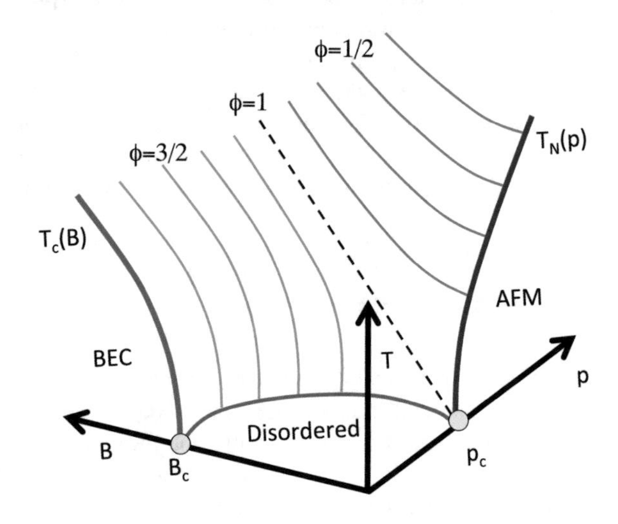

Fig. 7.2 Multiple universalities in the (p, B, T) phase diagram. Blue curves correspond to the BEC transition lines; here $p < p_c$ and the critical exponent is $\phi = 3/2$. Red curves correspond to the Néel transition lines; here $p > p_c$ and the critical exponent is $\phi = 1/2$. The dashed, black curve shows the critical pressure transition line, with critical exponent $\phi = 1$

The primary goal of the present work is to derive the evolution of the critical index ϕ across the phase diagram. Our secondary goal will be to explain the shortcomings of previous approaches to magnon Bose-condensation presented in the literature.

7.2 Methods

We work with the Lagrangian [1–3],

$$\mathscr{L}[\bar{\varphi}, \boldsymbol{B}] = \frac{1}{2}\left(\partial_t \bar{\varphi} - \bar{\varphi} \times \boldsymbol{B}\right)^2 + \frac{1}{2}c^2(\nabla\bar{\varphi})^2 + \frac{1}{2}m_0^2\bar{\varphi}^2 + \frac{1}{4}\alpha_0^2\bar{\varphi}^4. \tag{7.2}$$

The vector field $\vec{\varphi}$ describes staggered magnetisation, B is an external applied field, and for now we set $g\mu_B = 1$. We take the linear expansion $m_0^2(p) = \gamma^2(p_c - p)$, where $\gamma^2 > 0$ is a coefficient and p is the applied pressure. Presence of the $\boldsymbol{B}$-field explicitly breaks the $O(3)$ rotational symmetry down to an $O(2)$ rotational symmetry about the axis defined by $\boldsymbol{B}$. The three degenerate modes of the symmetric phase are Zeeman split such that the excitation gaps are $\Delta_\sigma = m_0 + \sigma B$, where $\sigma = 0, \pm 1$ are the polarisations (the projection of angular momentum on the direction of magnetic field). Hence only the $\sigma = -1$ mode becomes critical. Figure 7.3a depicts this situation. Upon magnetic field driven condensation, staggered magnetic order develops and lies in a plane perpendicular to the axis defined by the applied magnetic field $\boldsymbol{B}$. At zero temperature, the field induced QPT results from tuning $B > m_0 = \gamma\sqrt{p_c - p}$, this corresponds to the blue line in the (B, p) plane, Fig. 7.2. The condensate field is given by $\varphi_c^2 = \left(B^2 - m_0^2\right)/\alpha_0$. Within this phase, there exists one Goldstone mode, and two gapped modes. Of the gapped modes, one is an amplitude fluctuation or Higgs mode, while the other is a precession mode with rest energy set by the Larmor frequency, $g\mu_B B$, where g is the gyromagnetic factor and μ_B the Bohr magneton. The evolution of the excitation gaps through the magnetic field driven and pressure driven (at $B \neq 0$) QCPs are presented in Fig. 7.3.

7.2.1 Comparison with Standard BEC Effective Field Theory

The magnetic field changes the universality of the quantum phase transition—only one Goldstone mode is generated, as opposed to the $O(3)$ QPT which generates two Goldstone modes. Having just one critical mode and global $O(2)$ symmetry, this effective field theory belongs to the 3+1 dimensional BEC universality class [4]. Furthermore, based on this idea, one can eliminate the higher energy modes, and extract a *lower-energy* effective theory, namely the following,

$$\mathscr{L}[\bar{\varphi}_\perp, \boldsymbol{B}] \approx B\varphi_x\partial_t\varphi_y - B\varphi_y\partial_t\varphi_x + \frac{1}{2}c^2(\nabla\bar{\varphi}_\perp)^2 + \frac{1}{2}(m_0^2 - B^2)\bar{\varphi}_\perp^2 + \frac{1}{4}\alpha_0\bar{\varphi}_\perp^4 \,. \tag{7.3}$$

Here the second order time derivatives are ignored under the assumption $B \gg \omega$, and the φ_z mode has been dropped as it is non-critical (note $\boldsymbol{B} = B\hat{z}$). We call this the *standard BEC effective field theory*. The Bose-Einstein condensation (of magnons) in dimerised quantum antiferromagnets has been considered on the basis of such an effective theory in a number of theoretical works [5–7]. Importantly, this critical theory is non-relativistic; dynamical critical exponent $z = 2$, and hence effective dimensionality is $d + z = 5$. The theory now sits above the upper critical dimension $D_c = 4$, and therefore observables do not receive logarithmic corrections. In contrast, the original field theory (7.2) receives logarithmic corrections, although there is no associated asymptotic freedom as B acts as an infrared cutoff. Despite the absence of an asymptotically free point at the QCP, the logarithmic corrections in the presence

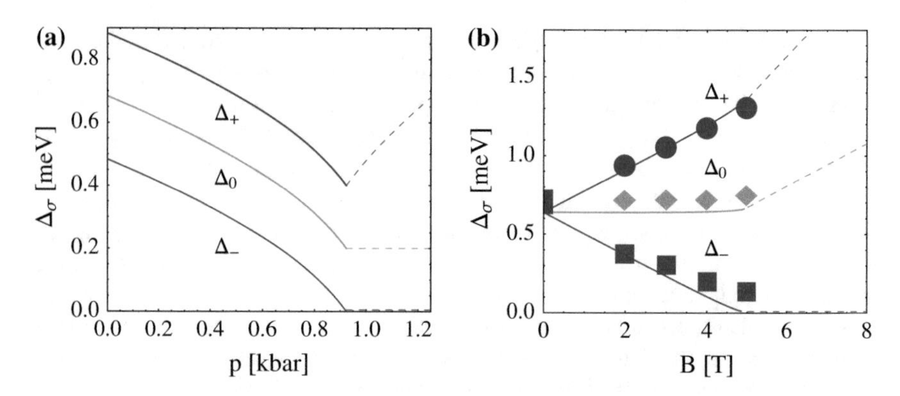

Fig. 7.3 Excitation gaps $\Delta_\sigma = \omega^\sigma_{k=0}$: (Left) pressure driven at fixed field $B = 0.2$ meV and $T = 0$. (Right) field driven at $p = 0$ kbar and $T = 1.5$ K. Solid lines are theoretical results derived in this chapter. Markers indicate experimental data for $TlCuCl_3$ [18, 19]

of a magnetic field will soon be seen to play an important role on the properties of the magnetic phase diagram, Fig. 7.2.

Under the mantra of effective QFT, many approaches to magnon Bose-condensation have relied on the *critical* Lagrangian (7.3), while certainly asymptotically correct (at vanishing energy), it has failed to describe real experimental data—the BEC critical index appeared different from theory. As such, the classification of magnetic field induced magnon condensation as belonging to the O(2) BEC universality has been an open question for ~20 years [5–16].

Moreover, it is widely believed that at $p < p_c$, $\phi = 3/2$ is the universal BEC exponent, which can be obtained from the scaling arguments on the dilute Bose gas [4, 17] or explicitly for magnon BEC [5, 6]. For a review see [7]. On the other hand, experiments on $TlCuCl_3$ and $KCuCl_3$ [8–13] and numerics [14] show $1.5 \leq \phi \lesssim 2.3$. With the exact value of ϕ depending crucially on which temperature range is used for fitting [7, 15]. By appealing to Eq. 7.2 instead of the approximate form Eq. 7.3, this chapter will explain why the index depends on the fitting range.

It is important now to comment on the validity of the critical theory (7.3). Only asymptotically close the the BEC quantum critical point can one expect this modified field theory (7.3) to provide a faithful representation of the physics, namely, when $B_c \gg T$. Physically, temperature acts as a bath of low energy modes. The discarding of modes, i.e. the precession and amplitude modes, may be justified in some limited region $B \gg \omega$ at $T = 0$, however, at finite T such modes can be readily excited (thermally) and as $T \sim B$ they become relevant degrees of freedom.

The effective theory (7.3) also has no prospect of describing decay channels of non-critical modes, including the Higgs modes; this will be the topic of Chap. 8.

To summarise, we claim that if one is interested in extended regions of the phase diagram, the influence of temperature and quasiparticle decay (non-critical modes), for which the full three-mode Lagrangian (7.2) is essential. Let us now proceed to the analysis of this field theory.

7.2.2 Renormalization

The parameters m_0 and α_0, do not include the renormalization due to quantum or thermal fluctuations. The fluctuations arise due to the interaction term $\frac{1}{4}\alpha_0\vec{\varphi}^4$ in Eq. (7.2). The one-loop renormalization procedure is explicitly performed in Appendix E. We denote the renormalized parameters $m_0^2 \to m_{\Lambda,\sigma}^2$ and $\alpha_0 \to \alpha_\Lambda$, which are now dependent on the energy scale Λ. The running coupling is given by

$$\alpha_\Lambda = \frac{\alpha_0}{1 + 11\alpha_0/(8\pi^2)\ln(\Lambda_0/\Lambda)}. \tag{7.4}$$

Specifically for the problem at hand, the coupling runs with scale $\Lambda = \max\{m_{\Lambda,\sigma}, B, T\}$. Accordingly, there is just a single point on the phase diagram at which all energy scales vanish $\Lambda \to 0$: the quantum critical point $(p_c, 0, 0)$, see Fig. 7.2. At this point the coupling runs to zero $\alpha_\Lambda \to 0$ (asymptotic freedom).

To present the renormalised mass $m_{\Lambda,\sigma}$, we first need to obtain the dispersions of the modes in the disordered phase. They follow from the Euler-Lagrange equations of (7.2),

$$\omega_k^\sigma = \sqrt{k^2 + m_{\Lambda,\sigma}^2} + \sigma B. \tag{7.5}$$

The explicit form for $m_{\Lambda,\sigma}^2 = m_{\Lambda,\sigma}^2(p, T, B)$ depends on the location within the phase diagram, and polarisation σ. Note that the σB term is not renormalised. All in all, calculations presented in Appendix E give,

$$\begin{aligned}
m_{\Lambda,\pm}^2 &= m_0^2\left[\frac{\alpha_\Lambda}{\alpha_0}\right]^{\frac{5}{11}} + \Sigma_T\ , \\
m_{\Lambda,0}^2 &= m_0^2\left[\frac{\alpha_\Lambda}{\alpha_0}\right]^{\frac{5}{11}} + \alpha_\Lambda \sum_k 1/\omega_k^0\{n(\omega_k^+) + n(\omega_k^-) + 3n(\omega_k^0)\}\ , \\
\Sigma_T &\equiv \alpha_\Lambda \sum_k 1/\omega_k^0\{2n(\omega_k^+) + 2n(\omega_k^-) + n(\omega_k^0)\}\ .
\end{aligned} \tag{7.6}$$

Here $n(\omega_k) = 1/(e^{\frac{\omega_k}{T}} - 1)$, and we introduce the function Σ_T for brevity.

In Fig. 7.3 we summarise the results for the evolution of the three mode gaps through the field and pressure quantum phase transitions, separately. The mode gaps, $\Delta_\sigma = \omega_{k=0}^\sigma$, are calculated from Eqs. (7.5) and (7.6). Explicit parameters correspond to those found in Chap. 2 and Ref. [20] for $TlCuCl_3$. Here we disregard the small easy-plane anisotropy seen in $TlCuCl_3$, which has been shown to have negligible influence on the critical properties [20], see also comment [21].

In a magnetic field, the condition of condensation follows from Eq. (7.5), $m_{\Lambda,\pm} - B_c = 0$. Using (7.6) this equation can be rewritten as,

$$\Sigma_T = B_c^2 - m_0^2 \left[\frac{\alpha_\Lambda}{\alpha_0}\right]^{\frac{5}{11}}, \tag{7.7}$$

which contains all information about the critical indices across the entire (p, B, T) phase diagram. It is our key theoretical result. Next we analyse this equation in the three qualitatively distinct regimes of the phase diagram.

7.3 Results and Discussion

We analyse three distinct cases: (I) Above the critical pressure $p > p_c$, here the critical temperature is given by the AFM/Néel temperature $T_c = T_N$; (II) exactly at the critical pressure, $p = p_c$; (III) below the critical pressure $p < p_c$, which we denote $T_c = T_{\text{BEC}}$.

7.3.1 *Case I*

Consider case (I), where $p > p_c$. In this case according to Eq. (7.1b) the Néel temperature varies in a weak magnetic field. To calculate Σ_T at $B \to 0$ we take the critical line dispersions $\omega_k^+ = \omega_k^- = \omega_k^0 = k$. Hence $\Sigma_T = \frac{5\alpha_\Lambda}{12} T^2$, where $T = T_{N0} + \delta T_N$, and T_{N0} denotes the Néel temperature in zero magnetic field. Hence using Eq. (7.7) we find,

$$\text{(I):} \quad \delta T_N = \frac{6}{5\alpha_\Lambda}\frac{B^2}{T_{N0}}, \qquad \text{at} \quad B \ll T_{N0}. \tag{7.8}$$

Hence for case (I) we conclude that the the critical index in Eq. (7.1b) is $\phi = 1/2$. This regime has never been considered before.

7.3.2 *Case II*

Consider case (II), where we now tune exactly to the quantum critical point, $p = p_c$, $T_{N0} = 0$. Again, to calculate Σ_T at $B \to 0$ we have to take the critical line dispersions $\omega_k^+ = \omega_k^- = \omega_k^0 = k$ and hence again $\Sigma_T = \frac{5\alpha_\Lambda}{12} T^2$. Substitution into (7.7) gives,

$$\text{(II):} \quad B_c = \sqrt{\frac{5\alpha_\Lambda}{12}}\, T, \qquad \text{at} \quad B_c \ll T. \tag{7.9}$$

The condition $B_c \ll T$ is satisfied at sufficiently low temperatures since the coupling constants decays logarithmically, $\alpha_\Lambda \propto 1/\ln\left(\frac{\Lambda_0}{T}\right)$. Hence in this case (II), the critical index of Eq. (7.1) is $\phi = 1$. We find that, in addition to the exponent, there is nontrivial logarithmic scaling. This regime has never been considered before.

7.3.3 *Case III*

Finally we consider the BEC case (III), where $p < p_c$. In this case only the $\omega_{\boldsymbol{k}}^-$ dispersion branch is critical, $\omega_{\boldsymbol{k}}^- \approx \frac{k^2}{2\Delta_0}$, where $\Delta_0 = B_0$ is the gap at $B = 0$. The other two branches are gapped. Calculation of Σ_T gives $\Sigma_T = \alpha_\Lambda \frac{\zeta(3/2)}{\pi\sqrt{2\pi}}\sqrt{\Delta_0}T^{3/2}$, where ζ is Riemann's ζ-function. Hence, using Eq. (7.7) we find,

$$\text{(III):}\quad \frac{\delta B_c}{\Delta_0} = \alpha_\Lambda \frac{\zeta(3/2)}{(2\pi)^{\frac{3}{2}}}\left(\frac{T}{\Delta_0}\right)^{3/2} \qquad \text{at} \quad \delta B_c \ll \Delta_0. \tag{7.10}$$

As expected the critical index in Eq. (7.1a) is $\phi = 3/2$. Such a power could have been derived from the approximate form of the Lagrangian (7.3). Except, by using the full form (7.2), we obtain multiplicative logarithmic corrections to this power law; this is due to the appearance of pre-factor α_Λ. Such multiplicative logarithmic corrections have not been considered before.

7.3.4 *Discussion*

We now discuss agreement between the asymptotic forms Eqs. (7.8), (7.9) and (7.10), the full solution (7.7) and the experimental data from [13, 22, 23]. To do so, we take the set of parameters describing $TlCuCl_3$, which were determined in Chap. 2

$$\begin{aligned} p_c &= 1.01 \text{ kbar},\ \gamma = 0.68 \text{ meV/kbar}^{1/2}, \\ \frac{\alpha_0}{8\pi} &= 0.23, \qquad \Lambda_0 = 1 \text{ meV}. \end{aligned} \tag{7.11}$$

Note, when fitting experimental data in Chap. 2 the thermal line-broadening had been accounted via $\omega = k \to \omega = \sqrt{k^2 + \xi^2 T^2}$, $\xi = 0.15$.

In Fig. 7.4 we illustrate Eq. (7.8) by dashed yellow line originating from $T_{N0} = 2.8$ K. The coupling constant is $\alpha_\Lambda/8\pi = \alpha_{T_{N0}}/8\pi = 0.107$. For comparison, the solid blue line originating from 2.8 K represents exact solution of Eq. (7.7) with coupling constant running along the line. We see that the asymptotic solution provides a faithful description over a large (B, T) parameter range. There is as yet no available experimental data for this regime.

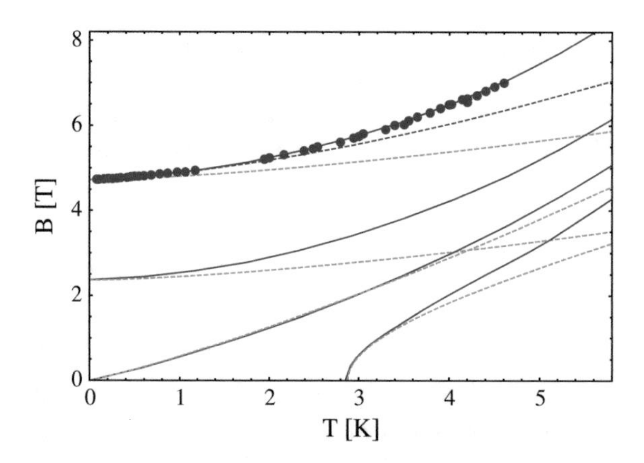

Fig. 7.4 Critical field vs temperature: Dashed yellow curves show solutions to scaling Eqs. (7.8), (7.9) and (7.10). Dashed maroon shows solution of (7.7) that accounts for thermal mixing of non-critical modes, but does not account for running coupling; coupling is at fixed value $\alpha_\Lambda \to \alpha_{\Delta_0} = 0.169 \times 8\pi$. Solid blue lines are the solution to (7.7) with full account of non-critical modes and logarithmic running coupling. Blue points are experimental data from [13, 22, 23]

We illustrate the asymptotic (7.9) by the dashed yellow line originating from $B = T = 0$ in Fig. 7.4. The solid blue line originating from the same point represents exact solution of Eq. (7.7). Once again, the asymptotic form is a faithful description and there is as yet no experimental data.

Finally, we consider the region of validity of Eq. (7.10). To do so we compare with both the full solution to (7.7) and with $TlCuCl_3$ data [13, 22, 23]. The value of the gap at $T = p = B = 0$ is $\Delta_0 = m_{\Lambda,\pm} = 0.64$ meV [24]. The BEC critical field for $T = p = 0$ is $B_0 = 4.73$ T [25]. Hence, we obtain the g-factor, $g = 2.35$ [21]. In Fig. 7.4 the dashed yellow line originating from $B_0 = 4.73$ T shows $B_{\rm BEC}$ versus T at $p = 0$ calculated with Eq. (7.10). The value of the coupling constant in this equation is obtained from Eqs. (7.4) and (7.11), $\alpha_\Lambda/(8\pi) = \alpha_{\Delta_0}/(8\pi) = 0.169$. Experimental data [13, 22, 23] are shown by circles. We see that Eq. (7.10) is valid only at $T \leq 1$ K.

There are two physical effects accounted in (7.7), but neglect in (7.10). These are (i) the influence of the non-critical (gapped) modes ω_k^+, ω_k^0, and (ii) the logarithmic running of α_Λ. To illustrate the importance of non-critical modes, the dashed maroon line originating from 4.73 T in Fig. 7.4 shows solution solution of Eq. (7.7) with account of all three modes, but with fixed coupling constant $\alpha_{\Delta_0}/(8\pi) = 0.169$. Finally, the solid blue line originating from 4.73 T shows solution of (7.7) with account of both (i) and (ii). Agreement with experiment is remarkable. And hence we conclude that by including the effects (i) and (ii), the present analysis resolves the long standing problem of the BEC critical exponent, which has been consistently reported at a higher value; $3/2 \leq \phi \leq 2.3$ [7–15]. We stress that there is no fitting in the theoretical curve. The set of parameters (7.11) was determined in Chap. 2 and Ref. [20] from data unrelated to magnetic field.

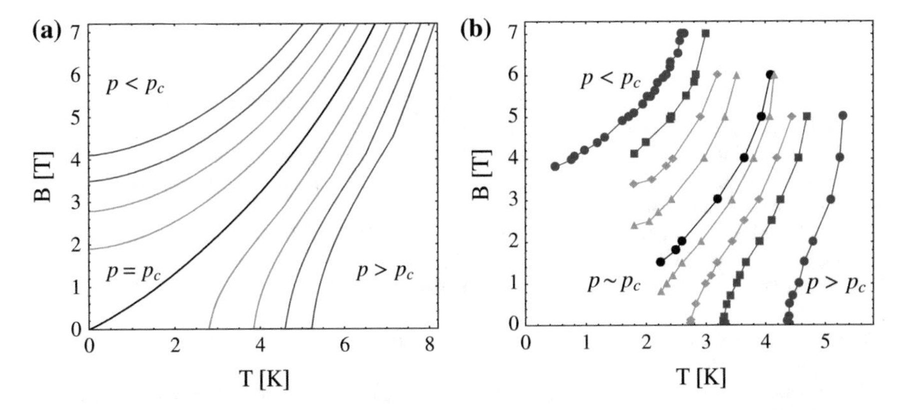

Fig. 7.5 Multiple universalities: Various curves show the critical field $B_c(T)$ at various pressures ranging $p < p_c$, $p = p_c$ to $p > p_c$. **a** Solutions to (7.7) with parameters for $TlCuCl_3$. **b** Data for quantum antiferromagnet $CsFeCl_3$ [16]

The existence of three critical exponents $\phi = 3/2$, 1 and 1/2, and even logarithmic corrections to these exponents, is a readily testable result and constitutes our most important prediction for experiment. There is as yet no experimental data on $TlCuCl_3$ in the regimes (I) and (II). However, there is experimental data on quantum antiferromagnet $CsFeCl_3$ which covers all three regimes. To make use of this data, in Fig. 7.5a we plot the predicted critical field in $TlCuCl_3$ vs temperature at various pressures. And for comparison in Fig. 7.5b we present a similar experimental plot for $CsFeCl_3$ [16]. For this comparison, we have not performed exact quantitative calculations (including all pre-factors) for $CsFeCl_3$. The existing data for this compound are not sufficient to perform analysis similar to Chap. 2 for $TlCuCl_3$. However, the data [16] shows clear qualitative support for the proposed *multiple universality* theory developed in this work.

7.4 Conclusion

In summary, we have employed a quantum field theoretic approach that allows access to the entire (p, B, T) phase diagram for 3D quantum antiferromagnets. Performing one-loop renormalization at non-zero temperatures, our approach allows us to uncover two new universality classes; we find two new critical exponents, as well as their corresponding logarithmic corrections. We also resolve a long standing problem relating to a third, known critical exponent in these systems. Our results show remarkable agreement with existing data on $TlCuCl_3$.

References

1. Fisher DS (1989) Universality, low-temperature properties, and finite-size scaling in quantum antiferromagnets. Phys Rev B 39:11783–11792
2. Sachdev S (2011) Quantum phase transitions. Cambridge University Press
3. Affleck I (1991) Bose condensation in quasi-one-dimensional antiferromagnets in strong fields. Phys Rev B 43:3215–3222
4. Fisher MPA, Weichman PB, Grinstein G, Fisher DS (1989) Boson localization and the superfluid-insulator transition. Phys Rev B 40:546–570
5. Giamarchi T, Tsvelik AM (1999) Coupled ladders in a magnetic field. Phys Rev B 59:11398–11407
6. Nikuni T, Oshikawa M, Oosawa A, Tanaka H (2000) Bose-Einstein condensation of dilute magnons in $TlCuCl_3$. Phys Rev Lett 84:5868–5871
7. Kawashima N (2005) Critical properties of condensation of field-induced triplet quasiparticles. J Phys Soc Jpn 74(Suppl):145–150
8. Shiramura W, Takatsu KI, Tanaka H, Kamishima K, Takahashi M, Mitamura H, Goto T (1997) High-field magnetization processes of double spin chain systems $KCuCl_3$ and $TlCuCl_3$. J Phys Soc Jpn 66(7):1900–1903
9. Kato T, Takatsu KI, Tanaka H, Shiramura W, Mori M, Nakajima K, Kakurai K (1998) Magnetic excitations in the spin gap system $KCuCl_3$. J Phys Soc Jpn 67(3):752–754
10. Oosawa A, Takamasu T, Tatani K, Abe H, Tsujii N, Suzuki O, Tanaka H, Kido G, Kindo K (2002) Field-induced magnetic ordering in the quantum spin system $KCuCl_3$. Phys Rev B 66:104405
11. Oosawa A, Ishii M, Tanaka H (1999) Field-induced three-dimensional magnetic ordering in the spin-gap system $TlCuCl_3$. J Phys Condens Matter 11(1):265
12. Cavadini N, Heigold G, Henggeler W, Furrer A, Güdel H-U, Krämer K, Mutka H (2001) Magnetic excitations in the quantum spin system $TlCuCl_3$. Phys Rev B 63:172414
13. Tanaka H, Oosawa A, Kato T, Uekusa H, Ohashi Y, Kakurai K, Hoser A (2001) Observation of field-induced transverse nel ordering in the spin gap system $TlCuCl_3$. J Phys Soc Jpn 70(4):939–942
14. Wessel S, Olshanii M, Haas S (2001) Field-induced magnetic order in quantum spin liquids. Phys Rev Lett 87:206407
15. Nohadani O, Wessel S, Normand B, Haas S (2004) Universal scaling at field-induced magnetic phase transitions. Phys Rev B 69:220402
16. Kurita N, Tanaka H (2016) Magnetic-field- and pressure-induced quantum phase transition in $CsFeCl_3$ proved via magnetization measurements. Phys Rev B 94:104409
17. Uzunov DI (1981) On the zero temperature critical behaviour of the nonideal Bose gas. Phys Lett A 87(1):11–14
18. Rüegg C, Cavadini N, Furrer A, Krämer K, Güdel HU, Vorderwisch P, Mutka H (2002) Spin dynamics in the high-field phase of quantum-critical $S = 1/2$ $TlCuCl_3$. Appl Phys A 74(1):s840–s842
19. Rüegg C, Cavadini N, Furrer A, Güdel H-U, Krämer K, Mutka H, Wildes A, Habicht K, Vorderwisch P (2003) Bose-Einstein condensation of the triplet states in the magnetic insulator $TlCuCl_3$. Nature 423(6935):62–65
20. Scammell HD, Sushkov OP (2015) Asymptotic freedom in quantum magnets. Phys Rev B 92:220401
21. Due to a small anisotropy, the g-factor slightly depends on the direction of magnetic field with respect to the crystal axes. We use BEC data with magnetic field directed as per experiment
22. Oosawa A, Katori HA, Tanaka H (2001) Specific heat study of the field-induced magnetic ordering in the spin-gap system $TlCuCl_3$. Phys Rev B 63:134416

23. Shindo Y, Tanaka H (2004) Localization of spin triplets in $Tl_{1-x}K_xCuCl_3$. J Phys Soc Jpn 73(10):2642–2645
24. Rüegg C, Normand B, Matsumoto M, Furrer A, McMorrow DF, Krämer KW, Güdel HU, Gvasaliya SN, Mutka H, Boehm M (2008) Quantum magnets under pressure: controlling elementary excitations in $TlCuCl_3$. Phys Rev Lett 100:205701
25. Yamada F, Ono T, Tanaka H, Misguich G, Oshikawa M, Sakakibara T (2008) Magnetic-field induced Bose-Einstein condensation of magnons and critical behavior in interacting spin dimer system $TlCuCl_3$. J Phys Soc Jpn 77(1):013701

Chapter 8
Prediction of Ultra-Narrow Higgs Resonance in Magnon Bose-Condensates

Abstract Higgs resonance modes in condensed matter systems are generally broad, meaning large decay widths/short relaxation times. This common feature has obscured and limited their observation to a select few systems. Contrary to this, the present work predicts that Higgs resonances in magnetic field induced, three-dimensional magnon Bose-condensates have vanishingly small decay widths. Crucially, our work demonstrates that an applied magnetic field acts as a direct tuning handle—controlling the strength of the coupling of Higgs to low energy modes, and hence the Higgs decay width. We calculate the evolution of the decay width under magnetic field for generic magnon Bose-condensates. Specifically for parameters relating to $TlCuCl_3$, we find an energy (Δ_H) to width (Γ_H) ratio $\Delta_H/\Gamma_H \sim 500$, making this predicted Higgs mode two orders of magnitude 'narrower' than for the same system without magnetic field.

8.1 Introduction

The Higgs mechanism, and associated Higgs modes, play a central role in modern physics. The mechanism is responsible for the mass generation of all observed particles in nature, and is the only known universal mechanism to do so. Higgs modes are a generic property of systems with a spontaneously broken continuous symmetry. This includes prominent condensed matter phenomena, superconductivity, Bose-condensation (BEC) and superfluidity, quantum magnetism, etc. as well as the Electroweak vacuum. Due to the ubiquity and importance of Higgs modes across many branches of physics, their detection has been an exciting, yet difficult, challenge. Certainly the discovery of the Electroweak Higgs boson [1, 2] meets both of these descriptions. Also attracting a great deal of attention, and proving to host their own difficulties, are the Higgs modes of condensed matter systems. They have been observed in the following settings, superfluid ^{3}He-B (1980) [3, 4], the charge density wave superconductor $NbSe_2$ (1981) [5, 6], three dimensional quantum antiferromagnet (AFM) $TlCuCl_3$ (2008) [42], superfluid ^{87}Rb atoms in an optical lattice (2012) [7], superconducting NbN (2013) [8, 9], and two dimensional quantum AFMs, Ca_2RuO_4 (2017) [10] and $C_9H_{18}N_2CuBr_4$ (2017) [11, 12]. Each setting offers unique insights

H. Scammell, *Interplay of Quantum and Statistical Fluctuations in Critical Quantum Matter*, Springer Theses,
https://doi.org/10.1007/978-3-319-97532-0_8

into the dynamics of Higgs modes and, in particular, the role played by symmetry, dimensionality, as well as the coupling to different degrees of freedom. Such factors are seen to have a dramatic influence on the dynamical properties and, ultimately, the observability of the Higgs modes.

A dimensionless parameter characterising the *quality* of the Higgs mode is the ratio of the mode energy over the decay width, $Q = \Delta_H/\Gamma_H$. For many interesting, symmetry-broken systems, the Higgs modes obtain $Q \sim 1$, implying such poor quality that the mode is unobservable. This is the case for the Higgs partners to the following low energy modes: the Higgs partner to spin waves in simple Heisenberg AFM; the partner to π-mesons (chiral symmetry breaking) [13]; the partners to sound in atomic BEC and in superfluid 4-He, etc. It is worth noting that in some cases it is possible to detect Higgs modes indirectly, even for low $Q \sim 1$. For example, in the case of superconductors $NbSe_2$ and NbN, observation requires either the presence of charge density wave order, or the implementation of out-of-equilibrium spectroscopy [14–16].

On the other hand, there are only very few systems exhibiting high quality factor Higgs modes. First, the quality factor of the fundamental 125 GeV Higgs boson in particle physics is $Q \approx 2 \times 10^4$ [17], and is the largest known in nature. Another remarkable scenario is the Higgs squashing modes of the ^{3}He-B order parameter, such modes have been shown to be long lived, with $Q \sim 10^4$ [3, 4, 18]. Note, if a discrete symmetry (instead of continuous) is broken, as for an Ising transition, then the Higgs mode is naturally long lived with a high quality factor. This scenario has been realised very recently [11, 12], where the Goldstone modes are gapped due to anisotropy, placing the transition in the Ising universality class.

We are concerned with the Higgs modes arising three dimensional quantum antiferromagnetic systems, such as $TlCuCl_3$ and similar, whereby the phase transition breaks a continuous symmetry. As a result the Higgs mode couples to gapless Goldstone modes and therefore generally has a low quality factor. For example, the pressure induced antiferromagnetic phase of $TlCuCl_3$ hosts the highest observed quality factor for this class of systems, with a mere $Q = 5$ [7]. The *high* quality factor is, in part, due to proximity to the quantum critical point (QCP) where the system becomes asymptotically free, as discussed in Chap. 2. A result related to asymptotic freedom was obtained in Ref. [46] in the context of quasi-one-dimensional chains. Ultimately, asymptotic freedom is a manifestation of the running coupling constant, α. For ease of presentation, in the present chapter we ignore the running of the coupling constant by setting $\alpha =$constant. To incorporate the running of α one can simply apply the results of Chap. 2; Eq. (2.2).

In this chapter we predict that the Higgs resonance in isotropic quantum antiferromagnetic systems can be made very narrow by application of an external magnetic field, obtaining a quality factor $Q \sim 500$. Moreover, with account of small anisotropy, as in the case of $TlCuCl_3$, and applying laboratory strength magnetic fields ($B \sim 6$ T), we obtain quality factors as high as the largest known in nature, $Q \sim 10^4$. The predicted resonance width is so narrow that it may be beyond the resolution of inelastic neutron scattering techniques, with meV resolution [7]. Instead measurement would require μeV resolution, for which neutron spin-echo technique is appropriate [19].

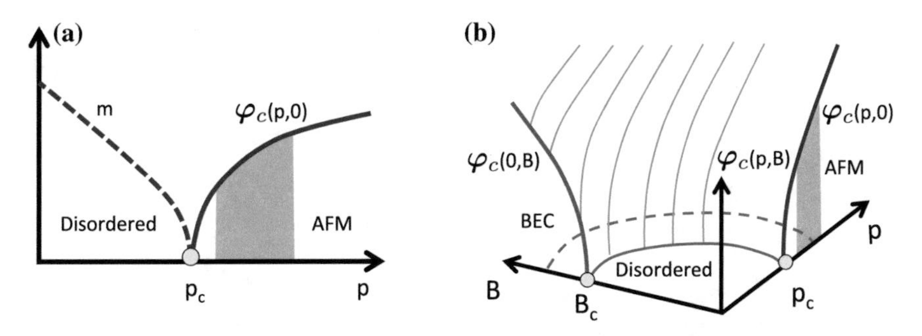

Fig. 8.1 Zero temperature phase diagram. **a** Magnon gap ($p < p_c$) and spontaneous staggered magnetisation ($p > p_c$) versus pressure at zero magnetic field, $B = 0$. **b** Pressure-Magnetic field phase diagram. The vertical axis shows spontaneous staggered magnetisation. The dashed line in the $B - p$ plane shows a contour connecting the simple AFM phase at $B = 0$ and the BEC phase at $p = 0$. The red band in both panels indicates the region where the width of the Higgs excitation has been measured

Moreover, Raman spectroscopy, which probes the scalar response channel, has been used to study Higgs modes of magnon-Bose condensates [20, 21]. Raman spectroscopy may therefore provide a suitable means to study the Higgs decay width [22].

Although magnon BEC have attracted immense experimental [23–30] and theoretical [31–36] interest over the past two decades, see reviews [37, 38], the issue of the Higgs magnon width in the BEC phase has not been addressed. Theoretically, the width in the usual AFM phase, i.e. not the BEC, was considered in [49] and also in recent Monte Carlo simulations [39].

In this work we address three dimensional (3D) quantum AFMs, having in mind $TlCuCl_3$ and similar. The zero temperature phase diagram of the system we consider is shown in Fig. 8.1, where Fig. 8.1a corresponds to a zero magnetic field slice of Fig. 8.1b. The quantum phase transition is driven by an external parameter, say pressure p. At $p > p_c$ the system is in the AFM phase, the order parameter $\varphi_c \neq 0$ is proportional to the staggered magnetisation. The region $p < p_c$ corresponds to the magnetically disordered phase, where the excitations—triplons, are gapped and are triply degenerate. We denote by m the gap in the triplon spectrum. BEC of magnons at $p < p_c$ can be driven by external magnetic field B. Evolution of the triplon gap under the magnetic field is shown in Fig. 8.2a. At weak field there is simple Zeeman splitting of the triple degenerate gapped triplon. At the critical value of the field B_c the lowest dispersion branch strikes zero. This is the BEC critical point. At a higher field the lowest branch remains gapless, this is the Goldstone mode of the magnon BEC. Gaps in the middle branch (z-mode) and the top branch (Higgs mode) continue to evolve with magnetic field. The zero temperature $B - p$ phase diagram is presented in Fig. 8.1b where the vertical axis shows the order parameter. The diagram clearly indicates that the AFM phase at $B = 0$, $p > p_c$ is continuously connected with the BEC phase at $p = 0$, $B > B_c$ [41]. Evolution of excitation gaps with magnetic field at zero pressure, $p = 0$, is shown in Fig. 8.2a. Evolution of excitations gaps along

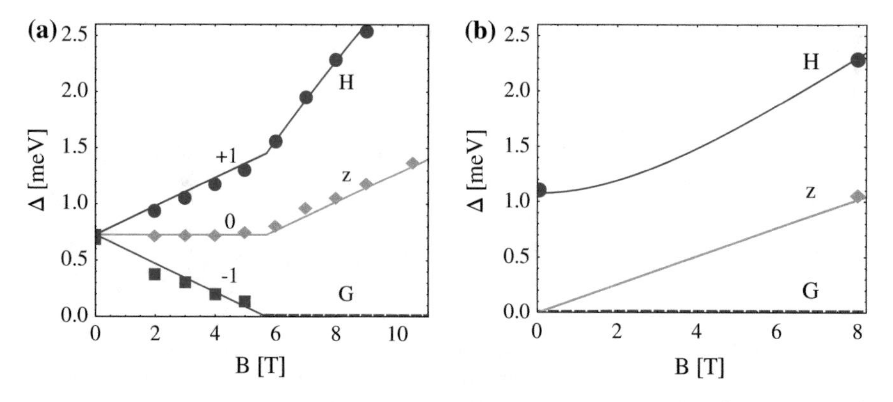

Fig. 8.2 Evolution of excitation gaps through the BEC phase transition. Markers indicate experimental data for $TlCuCl_3$ [43]. Solid lines are our theoretical results. **a** Gaps versus B-field at zero pressure. **b** Gaps versus B-field along the dashed line in Fig. 8.1b connecting AFM and BEC phases, described by $p(B) = p_0 - B^2/\gamma^2$ with $p_0 = 2.26$ kbar. Letters H, z, and G indicate Higgs, z- and Goldstone modes in the ordered phase. While numbers $\{-1, 0, +1\}$ refer to Zeeman split modes in the disordered phase, discussed in text. Experimental points are taken from [7, 43]

the dashed contour in the $B - p$ plane in Fig. 8.1b is shown in Fig. 8.2b. The contour corresponds to $p(B) = p_0 - B^2/\gamma^2$, where $p_0 = 2.26$ kbar is selected to correspond to experimental data [7, 43], and the parameter γ was introduced in Chap. 2, and is discussed again below. Markers in Fig. 8.2 indicate experimental data for $TlCuCl_3$ [7, 43] and solid lines represent theory described below.

8.2 Method: Isotropic Systems

The systems analysed here are close to quantum criticality, where usual spin-wave or triplon techniques are insufficient. Instead, the present analysis employs quantum field theory. The σ-model-type effective Lagrangian of the system reads [44–46],

$$\mathscr{L}[\vec{\varphi}] = \frac{1}{2}(\partial_t\vec{\varphi} - \vec{\varphi} \times \vec{B})^2 - \frac{1}{2}(\vec{\nabla}\vec{\varphi})^2 - \frac{1}{2}m^2\vec{\varphi}^{\,2} - \frac{1}{4}\alpha\vec{\varphi}^{\,4}. \tag{8.1}$$

Here $\vec{\varphi}$ is a real vector field describing AFM magnons, $m^2 = \gamma^2(p_c - p)$ is the pressure dependent effective mass (γ is a coefficient), and α is the coupling constant. In Eq. (8.1) we set the magnetic moment and the magnon speed equal to unity $g\mu_B = c = 1$. Of course, when comparing with experimental data these quantities have to be restored. Quantum fluctuations renormalise values of m^2 and α. The effect of renormalisation is well understood —the bare values of m and α are to be replaced by logarithmically renormalised values, $m \to m_R(\Lambda)$, $\alpha \to \alpha_R(\Lambda)$. In our analysis, the renormalisation scale, Λ, is equal to the Higgs gap. We will not discuss logarithmic renormalisation any further. We also strictly consider $T = 0$.

In the spontaneously symmetry-broken BEC phase, the classical expectation value immediately follows from (8.1), $\varphi_c^2 = (B^2 - m^2)/\alpha > 0$. Hence the field is $\vec{\varphi} = (\varphi_c + \sigma, \pi_y, \pi_z)$. The magnetic field defines the z-axis and we choose the x-axis to be directed along the spontaneous magnetisation, φ_c. Equation (8.1) rewritten in terms of dynamic fields, σ, π_y, and π_z, reads

$$\begin{aligned}
\mathscr{L} &= \mathscr{L}_2 + \mathscr{L}_3 + \mathscr{L}_4 \,, \\
\mathscr{L}_2 &= \frac{1}{2}[\dot{\sigma}^2 + \dot{\pi}_y^2 + \dot{\pi}_z^2] + B[\sigma\dot{\pi}_y - \dot{\sigma}\pi_y] \\
&\quad - \frac{1}{2}[\nabla\sigma^2 + \nabla\pi_y^2 + \nabla\pi_z^2] - (B^2 - m^2)\sigma^2 - \frac{1}{2}B^2\pi_z^2 \,, \\
\mathscr{L}_3 &= -\alpha\varphi_c\sigma(\sigma^2 + \pi_y^2 + \pi_z^2) \,, \\
\mathscr{L}_4 &= -\frac{\alpha}{4}(\sigma^2 + \pi_y^2 + \pi_z^2)^2 \,.
\end{aligned} \tag{8.2}$$

Fourier transforming the fields, $\sigma, \pi \propto e^{-i\omega t + i\boldsymbol{k}\cdot\boldsymbol{r}}$, and using the relation $\frac{1}{2}\varphi^T G^{-1}\varphi = \mathscr{L}_2$, we obtain the matrix Greens function

$$G^{-1} = \begin{pmatrix} \omega^2 - \boldsymbol{k}^2 - 2(B^2 - m^2) & 2iB\omega & 0 \\ -2iB\omega & \omega^2 - \boldsymbol{k}^2 & 0 \\ 0 & 0 & \omega^2 - \boldsymbol{k}^2 - B^2 \end{pmatrix} . \tag{8.3}$$

Evaluating $|G^{-1}| = 0$ results in the following dispersions for Higgs, Goldstone, and z-modes,

$$\begin{aligned}
\omega_{\boldsymbol{k}}^H &= \sqrt{\boldsymbol{k}^2 + 3B^2 - m^2 + \sqrt{4B^2\boldsymbol{k}^2 + (3B^2 - m^2)^2}} \,, \\
\omega_{\boldsymbol{k}}^G &= \sqrt{\boldsymbol{k}^2 + 3B^2 - m^2 - \sqrt{4B^2\boldsymbol{k}^2 + (3B^2 - m^2)^2}} \,, \\
\omega_{\boldsymbol{k}}^z &= \sqrt{\boldsymbol{k}^2 + B^2} \,.
\end{aligned} \tag{8.4}$$

Theoretical values of $\omega_{\boldsymbol{k}=0}$, given in Eq. (8.4), are plotted in Fig. 8.2 by solid lines [47]. The dispersions (8.4) have been obtained in the literature, see e.g. [46, 48], and do not represent a major conclusion of this chapter. Note that (8.4) is valid in the spontaneously broken phase, $B > B_c$. In the disordered phase, at $B < B_c$, there is simple Zeeman splitting of triplon dispersions, $\omega_{\boldsymbol{k}}^{(l)} = \sqrt{k^2 + m^2} + lB$, where $l = 0, \pm 1$.

Plots of dispersions (8.4) versus momentum for two values of magnetic field, $B = B_c$ and $B = 1.1B_c$ are presented in Fig. 8.3a. At the critical point, $B = B_c$, the Goldstone mode is quadratic in momentum at $k \to 0$. At $B > B_c$ the Goldstone mode is linear at $k \to 0$, yet with a non-linear bend at $k \sim B - B_c$.

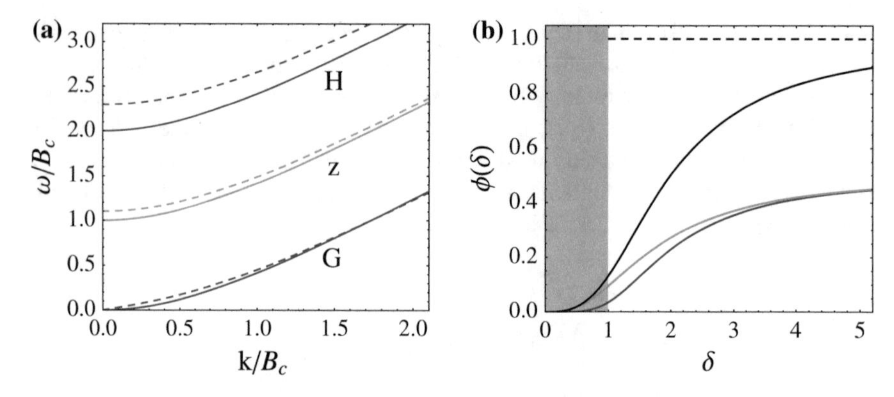

Fig. 8.3 **a** Dispersions of Higgs (H) z- and Goldstone (G) modes for $B = B_c$ (solid lines) and $B = 1.1B_c$ (dashed lines). **b** Scaling functions: $\phi_G(\delta)$ (maroon line) and $\phi_z(\delta)$ (yellow line). The parameter $\delta^2 = 1 - m^2/B^2$. Pink shaded region indicates BEC phase, $\delta < 1$. The AFM phase at $B = 0$ corresponds to $\delta = \infty$. The dashed black line represents the AFM limit of $\phi(\delta \to \infty)$

In principle, the decay amplitude of the Higgs mode is given by $\mathscr{L}_3$ and $\mathscr{L}_4$ in Eq. (8.2). However, in the BEC phase there is an important physical and technical complication that originates from the Berry phase term $B[\sigma\dot{\pi}_y - \dot{\sigma}\pi_y]$ in $\mathscr{L}_2$, which is not present in the usual AFM phase. This term introduces off-diagonal kinetic terms into the Lagrangian (or equivalently, into the Greens function (8.3)), which cause mixing, or hybridisation, between the σ and π_y modes. In a standard field theory or spin-wave theory without off-diagonal kinetic terms, the field operator is represented as a combination of corresponding creation and annihilation operators, $\sigma \propto \sum_k [a_k e^{i\omega_k t - i\boldsymbol{k}\cdot\boldsymbol{x}} + a_k^\dagger e^{-i\omega_k t + i\boldsymbol{k}\cdot\boldsymbol{x}}]$. This remains valid in the AFM phase of the present work. However, owing to the Berry phase, this is not valid in the BEC phase. Below we elaborate how to correctly represent the field operators, σ, π_y and π_z, in terms of creation/annihilation operators—it is important for understanding the results of the chapter.

Let $a_k/a_k^\dagger$, $b_k/b_k^\dagger$, and $c_k/c_k^\dagger$ be the annihilation/creation operators of the Higgs, Goldstone, and z- modes. Accordingly, the Hamiltonian reads,

$$H = \sum_k \left[\omega_k^H a_k^\dagger a_k + \omega_k^G b_k^\dagger b_k + \omega_k^z c_k^\dagger c_k\right] + \text{const.} \tag{8.5}$$

It is shown in Appendix F that the field operators are to be expressed in terms of the creation and annihilation operators in the following way,

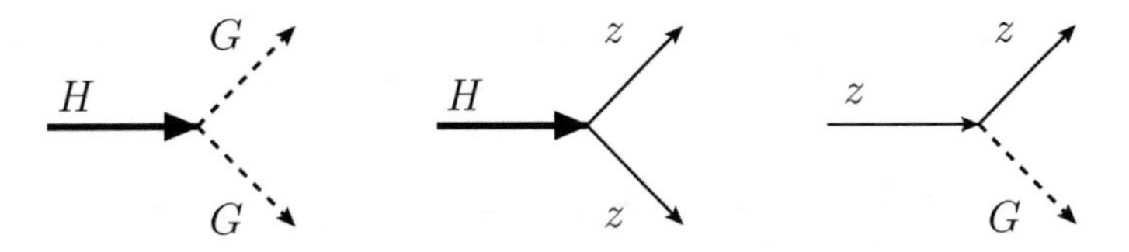

Fig. 8.4 Diagrams for the Higgs and z-mode decay channels

$$
\begin{aligned}
\sigma(\boldsymbol{x}, t) &= \sum_{\boldsymbol{k}} \Big\{ \mathscr{A}_{H,\boldsymbol{k}} [a_{\boldsymbol{k}} e^{i\omega_{\boldsymbol{k}}^{H} t - i\boldsymbol{k}\cdot\boldsymbol{x}} + a_{\boldsymbol{k}}^{\dagger} e^{-i\omega_{\boldsymbol{k}}^{H} t + i\boldsymbol{k}\cdot\boldsymbol{x}}] \\
&\qquad + \mathscr{A}_{G,\boldsymbol{k}} [b_{\boldsymbol{k}} e^{i\omega_{\boldsymbol{k}}^{G} t - i\boldsymbol{k}\cdot\boldsymbol{x}} + b_{\boldsymbol{k}}^{\dagger} e^{-i\omega_{\boldsymbol{k}}^{G} t + i\boldsymbol{k}\cdot\boldsymbol{x}}] \Big\} , \\
\pi_y(\boldsymbol{x}, t) &= \sum_{\boldsymbol{k}} \Big\{ \mathscr{B}_{H,\boldsymbol{k}} [a_{\boldsymbol{k}} e^{i\omega_{\boldsymbol{k}}^{H} t - i\boldsymbol{k}\cdot\boldsymbol{x}} - a_{\boldsymbol{k}}^{\dagger} e^{-i\omega_{\boldsymbol{k}}^{\pi} t + i\boldsymbol{k}\cdot\boldsymbol{x}}] \\
&\qquad + \mathscr{B}_{G,\boldsymbol{k}} [b_{\boldsymbol{k}} e^{i\omega_{\boldsymbol{k}}^{G} t - i\boldsymbol{k}\cdot\boldsymbol{x}} - b_{\boldsymbol{k}}^{\dagger} e^{-i\omega_{\boldsymbol{k}}^{G} t + i\boldsymbol{k}\cdot\boldsymbol{x}}] \Big\} , \\
\pi_z(\boldsymbol{x}, t) &= \sum_{\boldsymbol{k}} \frac{1}{\sqrt{2\omega_{\boldsymbol{k}}^{z}}} [c_{\boldsymbol{k}} e^{i\omega_{\boldsymbol{k}}^{z} t - i\boldsymbol{k}\cdot\boldsymbol{x}} + c_{\boldsymbol{k}}^{\dagger} e^{-i\omega_{\boldsymbol{k}}^{z} t + i\boldsymbol{k}\cdot\boldsymbol{x}}] ,
\end{aligned}
\tag{8.6}
$$

where, denoting $\alpha = H, G$, the coefficients are,

$$
\begin{aligned}
\mathscr{A}_{\alpha,\boldsymbol{k}} &= \sqrt{\frac{\omega_{\boldsymbol{k}}^{\alpha} B^2}{(B^2 + m^2)(\omega_{\boldsymbol{k}}^{\alpha})^2 + (3B^2 - m^2)(2B^2 - 2m^2 + \boldsymbol{k}^2)}} \\
\mathscr{B}_{\alpha,\boldsymbol{k}} &= D_{\alpha,\boldsymbol{k}} \mathscr{A}_{\alpha,\boldsymbol{k}} , \quad \mathscr{D}_{\alpha,\boldsymbol{k}} = \frac{-2i\omega_{\boldsymbol{k}}^{\alpha} B}{(\omega_{\boldsymbol{k}}^{\alpha})^2 - \boldsymbol{k}^2} .
\end{aligned}
\tag{8.7}
$$

We draw the readers attention to the minus signs in the canonical representation of $\pi_y(\boldsymbol{x}, t)$ (8.6), these signs are responsible for cancellation with the scattering matrix elements (8.8), and ultimately the narrowness of the Higgs mode.

In a conventional BEC, the Higgs mode is longitudinal and the Goldstone mode is transverse. As demonstrated in Eqs. (8.6), this is not true for the magnon BEC. Both longitudinal and transverse waves are linear combinations of Higgs and Goldstone excitations. The bending of the Goldstone dispersion at $k \sim B - B_c$, seen in Fig. 8.3a, is a direct manifestation of this hybridisation.

From the interaction term $\mathscr{L}_3$ (8.2) and applying (8.6), we conclude that Higgs can decay into two Goldstone excitations and into two z-excitations. This is shown diagrammatically in Fig. 8.4. The decay matrix elements follow from Eqs. (8.2), (8.6),

$$
\begin{aligned}
\mathscr{M}_{H\to GG} &= 2\alpha\varphi_c \mathscr{A}_{H,\boldsymbol{k}_0} \mathscr{A}_{G,\boldsymbol{k}_1} \mathscr{A}_{G,\boldsymbol{k}_2} \left\{ 3 + \mathscr{D}_{G,\boldsymbol{k}_1} \mathscr{D}_{G,\boldsymbol{k}_2} - \mathscr{D}_{G,\boldsymbol{k}_1} \mathscr{D}_{H,\boldsymbol{k}_0} - \mathscr{D}_{G,\boldsymbol{k}_2} \mathscr{D}_{H,\boldsymbol{k}_0} \right\} , \\
\mathscr{M}_{H\to zz} &= 2\alpha\varphi_c \frac{\mathscr{A}_{H,\boldsymbol{k}_0}}{\sqrt{2\omega_{\boldsymbol{k}_1}^{z} 2\omega_{\boldsymbol{k}_2}^{z}}} .
\end{aligned}
\tag{8.8}
$$

Here we denote the momentum of initial Higgs by $\boldsymbol{k}_0$ and momenta of final particles by $\boldsymbol{k}_1$ and $\boldsymbol{k}_2$. The decay width is given by Fermi's golden rule,

$$\Gamma = \frac{1}{2}(2\pi)^4 \int \frac{d^3k_1}{(2\pi)^3}\frac{d^3k_2}{(2\pi)^3}|\mathcal{M}|^2\delta(\omega_{\mathbf{k}_0}-\omega_{\mathbf{k}_1}-\omega_{\mathbf{k}_2})\delta(\boldsymbol{k}_0-\boldsymbol{k}_1-\boldsymbol{k}_2)\ , \quad (8.9)$$

where the coefficient 1/2 stands to avoid double counting of final bosonic states. A direct integration gives the decay widths. For clarity, and to avoid lengthy formulas, we present here only the partial widths in the rest frame, $\boldsymbol{k}_0 = 0$, $\omega^H = \sqrt{2(3B^2 - m^2)}$,

$$\frac{\Gamma_{H\to GG}}{\omega^H} = \frac{\alpha_0}{8\pi}\phi_G(\delta)\ , \quad (8.10)$$
$$\frac{\Gamma_{H\to zz}}{\omega^H} = \frac{\alpha_0}{8\pi}\phi_z(\delta)\ .$$

Here the scaling functions ϕ_G and ϕ_z depend on the parameter, $\delta^2 = \alpha\varphi_c^2/B^2 = 1 - m^2/B^2$,

$$\phi_G(\delta) = \frac{\delta^2}{2}\left(3 - \frac{2(2+\delta^2) - 4(\delta^2 - \sqrt{\delta^4+2\delta^2+4})}{(\delta^2-\sqrt{\delta^4+2\delta^2+4})^2}\right)^2$$
$$\times \frac{(2+\sqrt{\delta^4+2\delta^2+4})\sqrt{2-\delta^2+2\sqrt{\delta^4+2\delta^2+4}}}{(2+\delta^2)^{3/2}\sqrt{\delta^4+2\delta^2+4}(2+\delta^2+\sqrt{\delta^4+2\delta^2+4})^2}\ ,$$
$$\phi_z(\delta) = \frac{\delta^3/2}{(2+\delta^2)^{3/2}}\ . \quad (8.11)$$

The parameter, δ, ranges from $0 \le \delta \le \infty$, such that $\delta = 0$ at a BEC critical point, where $p < p_c$, $m^2 > 0$, and $B = m$, while $\delta = \infty$ in the AFM phase, where $p > p_c$, $m^2 < 0$, and $B = 0$. Asymptotics of the scaling functions are, $\phi_G(\infty) = \phi_z(\infty) = 1/2$, $\phi_G(\delta \to 0) \propto \delta^6$, $\phi_z(\delta \to 0) \propto \delta^3$. Plots of $\phi_G(\delta)$ and $\phi_z(\delta)$ are presented in Fig. 8.3b. A very strong suppression at small δ is evident, and this constitutes our main physical result.

8.2.1 *Q*-Factor Analysis

In the AFM phase at $B = 0$, $\delta = \infty$, the total width, $\Gamma = \Gamma_{H\to GG} + \Gamma_{H\to zz}$, has been already measured [7] and calculated [39, 49]. In this regime, Eq. (8.10) gives $\Gamma/\omega^H = \alpha/8\pi$, which is consistent with previous work [7, 39, 49]. The most interesting prediction of Eq. (8.10) is the dramatic suppression of the width at $\delta < \infty$ and especially in the BEC phase ($\delta < 1$). Explicitly for $TlCuCl_3$, taking $p = 0$ and $B = 1.1B_c \approx 6.4$ T, the Higgs gap is $\omega^H \approx 1.6$ meV and the quality factor $Q = \omega^H/\Gamma \approx$

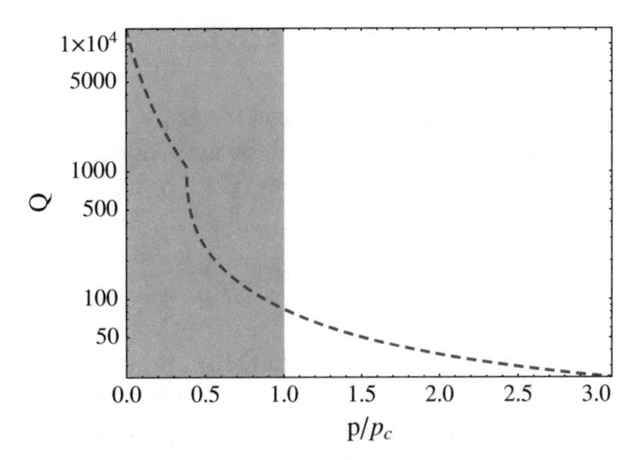

Fig. 8.5 Quality factor from Eq. (8.15) in the presence of spin-orbit anisotropy, evaluated with parameters relating to $TlCuCl_3$ and magnetic field $B = 2m_a$. Pink shaded region indicates pure BEC phase

500, predicted by Eq. (8.10), is much higher than that in the AFM phase, $Q \approx 5$. Even at the BEC border line, $p = p_c$, $\delta = 0$, the quality factor is $Q \approx 50$.

8.3 Method: Anisotropic Systems

For three dimensional quantum antiferromagnet $TlCuCl_3$, there exist a small easy plane anisotropy. We now analyse the influence of the anisotropy on the decay line width of the Higgs mode. If magnetic field is applied perpendicular to the easy plane, then we make the following adjustment to the Lagrangian (8.2),

$$\mathcal{L}_a[\varphi_c + \sigma, \pi_y, \pi_z] = \mathcal{L}[\varphi_c + \sigma, \pi_y, \pi_z] - \frac{1}{2} m_a^2 \pi_z^2 \,. \tag{8.12}$$

The additional term acts to further gap the z-mode, where $m_a \approx 0.38$ meV is known from experiment [7]. The Goldstone and Higgs dispersions (8.4) remain unchanged, however the z-mode dispersion becomes,

$$\tilde{\omega}_{\boldsymbol{k}}^z = \sqrt{\boldsymbol{k}^2 + B^2 + m_a^2} \,. \tag{8.13}$$

The partial decay into Goldstone modes, $\Gamma_{H\to GG}$, remains unaffected, but the partial decay into z-modes, $\Gamma_{H\to zz}$, is now influenced by the anisotropy as follows,

$$\begin{aligned} \frac{\tilde{\Gamma}_{H\to zz}}{\omega^H} &= \frac{1}{2} \frac{\alpha_0}{8\pi} \tilde{\phi}_2(\delta, a) \\ \tilde{\phi}_2(\delta, a) &= \frac{\delta^2}{(2+\delta^2)^{3/2}} \sqrt{\delta^2 - 2a^2}\,\Theta(\delta^2 - 2a^2) \,. \end{aligned} \tag{8.14}$$

8.3.1 Q-Factor Analysis

There are now two scaling parameters, $\delta^2 = 1 - m^2/B^2$ and $a = m_a/B$, which makes presentation of the analysis more difficult. So to finish this section, we just wish to present the quality factor, Q,

$$Q = \frac{\omega^H}{\Gamma_{H\to GG} + \tilde{\Gamma}_{H\to zz}} \,. \tag{8.15}$$

Here the partial decay widths, $\Gamma_{H\to GG}$ and $\tilde{\Gamma}_{H\to zz}$, are taken from Eqs. (8.10) and (8.14), respectively. We apply fitting parameters derived in Chap. 2 for $TlCuCl_3$, and also select a magnetic field strength, $B = 2m_a$. These parameters allow Q to be directly evaluated in terms of pressure, p. The result it plotted in Fig. 8.5. A step function non-analyticity appears due to the sudden threshold opening/closing of the decay channel into the anisotropic mode.

The anisotropy acts to dramatically enhance the quality factor. We therefore suggest scattering experiments in the parameter range $1 < p/p_c < 0.5$ and $B \approx 2m_a \approx$ 6 T. Importantly, this parameter range is experimentally realisable within current techniques [7].

8.4 Discussion

We remind the reader that the present analysis has neglected the logarithmic running of the coupling constant α, as obtained in Chap. 2; Eq. (2.2). Instead, the primary focus of this chapter was to demonstrate that an applied magnetic field acts as a direct tuning handle—controlling the strength of the coupling of Higgs to low energy modes, and hence the Higgs decay width. We now discuss how the running α influences the results obtained in this chapter. For the case of the $B = 0$, Eq. (8.10) gives $\Gamma/\omega^H = \alpha/8\pi$, which was also obtained in Chap. 3; see Eq. (3.21) and corresponding Fig. 3.5b. In Chap. 3 the running coupling was explicitly taken into account, and we see from Fig. 3.5b that precisely at the quantum critical point, the ratio $\Gamma/\omega^H = \alpha/8\pi \to 0$ and hence the Q-factor is infinite. The vanishing of the decay linewidth is a manifestation of asymptotic freedom. We note, however, the vanishing of the ratio $\Gamma/\omega^H = \alpha/8\pi \to 0$ occurs only in the very near vicinity of the QCP, see Fig. 3.5b, and that for any reasonable value of detuning away from the QCP the ratio $\Gamma/\omega^H = \alpha/8\pi \approx$ constant—since it has a logarithmically-shallow slope. We point out that a similar result was obtained in Ref. [46] in the context of quasi-one-dimensional chains. Now, for the case $B \neq 0$, the running coupling constant α does not vanish at the QCP—there is no asymptotic freedom. Hence, we can safely take α equal to a non-zero constant.

It is interesting to note that the z-mode is even more narrow than the Higgs. The dominant decay channel is via emission of the Goldstone excitation, $z \to z + G$,

Fig. 8.4. Moreover, the decay is possible only if the speed of z is higher than the speed of Goldstone excitation, making this a magnetic analog of Cherenkov radiation. Note, double Goldstone emission is also possible, $z \to z + G + G$, but with even lower amplitude.

Due to $\mathcal{L}_4$ in Eq. (8.2), the Higgs mode can also decay into three Goldstones, $H \to G + G + G$. However, the probability of this decay mode is much smaller than $H \to G + G$ considered above due to a reduced phase space and being at next-order in perturbative coupling α. At non-zero temperature Raman processes become possible, $H + G \to H + G$, $z + G \to z + G$. The corresponding broadening can be calculated using the developed technique supplemented with appropriate Bose-occupation factors. However, at low T, the Raman broadening is small due to Bose-occupation factors, and therefore we do not consider it here.

Comparing with real compounds one often has to account for weak spin-orbit anisotropy; we explicitly treat this scenario. The anisotropy slightly changes one or more mode dispersion. Which mode is affected depends on the orientation of the magnetic field. If the magnetic field is oriented such that the anisotropy only shows up as an additional gapping of the z-mode, the Higgs partial decay into the z-mode will be further reduced due to phase space, strengthening the present conclusions. Finally, it is worth noting that in the BEC phase the width becomes so narrow that the decay into two phonons may be comparable with purely magnetic decay mechanisms considered here.

In conclusion, we predict that Higgs modes in a magnon Bose-condensate phase of 3D quantum magnets are ultra-narrow, i.e. have vanishingly small decay width. We demonstrate that the Higgs mode in the isotropic Bose-condensate phase can be tuned to have a decay width two orders of magnitude smaller than the corresponding mode in the antiferromagnetic phase. An essential feature of the Bose-condensate is a Berry phase contribution that causes the collective modes; Higgs and Goldstone, to appear as the hybridisation of longitudinal and transverse excitations of the condensate order parameter. This hybridisation plays a key role in *narrowing* the Higgs mode. Moreover, we calculate dispersions of all collective excitations in the magnon Bose-condensate phase and find that hybridisation also manifests itself as a bending of the dispersion branches, thus providing further experimental tests of the scenario posed here.

References

1. CMS collaboration (1982) Observation of a new boson at a mass of 125 GeV with the CMS experiment at the LHC. Phys Lett B 716(1):30–61
2. ATLAS collaboration (2012) Observation of a new particle in the search for the standard model Higgs boson with the ATLAS detector at the LHC. Phys Lett B 716(1):1–29
3. Giannetta RW, Ahonen A, Polturak E, Saunders J, Zeise EK, Richardson RC, Lee DM (1980) Observation of a new sound-attenuation peak in superfluid ^{3}He $- b$. Phys Rev Lett 45:262–265
4. Mast DB, Sarma BK, Owers-Bradley JR, Calder ID, Ketterson JB, Halperin WP (1980) Measurements of high-frequency sound propagation in ^{3}He-B. Phy Rev Lett 45:266–269

5. Littlewood PB, Varma CM (1981) Gauge-invariant theory of the dynamical interaction of charge density waves and superconductivity. Phys Rev Lett 47:811–814
6. Littlewood PB, Varma CM (1982) Amplitude collective modes in superconductors and their coupling to charge-density waves. Phys Rev B 26:4883–4893
7. Endres M, Fukuhara T, Pekker D, Cheneau M, Schauß P, Gross C, Demler E, Kuhr S, Bloch I (2012) The Higgs amplitude mode at the two-dimensional superfluid/Mott insulator transition. Nature 487(7408):454–458
8. Matsunaga R, Hamada YI, Makise K, Uzawa Y, Terai H, Wang Z, Shimano R (2013) Higgs amplitude mode in the BCS superconductors $Nb_{1-x}Ti_xN$ induced by terahertz pulse excitation. Phys Rev Lett 111:057002
9. Matsunaga R, Tsuji N, Fujita H, Sugioka A, Makise K, Uzawa Y, Terai H, Wang Z, Aoki H, Shimano R (2014) Light-induced collective pseudospin precession resonating with Higgsmode in a superconductor. Science 345(6201):1145–1149
10. Jain A, Krautloher M, Porras J, Ryu GH, Chen DP, Abernathy DL, Park JT, Ivanov A, Chaloupka J, Khaliullin G, Keimer B, Kim BJ (2017) Higgs mode and its decay in a two-dimensional antiferromagnet. Nat Phys (advance online publication)
11. Qiu Y, Chen W, Gentile TR, Watson S, Awwadi FF, Turnbull MM, Dissanayake SE, Agrawal H, Toft-Petersen R, Klemke B, Coester K, Schmidt KP, Tennant DA, Hong T, Matsumoto M (2017). Direct observation of the higgs amplitude mode in a two-dimensional quantum antiferromagnet near the quantum critical point. arXiv:1705.06172
12. Hong T, Qiu Y, Matsumoto M, Tennant DA, Coester K, Schmidt KP, Awwadi FF, Turnbull MM, Agrawal H, Chernyshev AL (2017) Field induced spontaneous quasiparticle decay and renormalization of quasiparticle dispersion in a quantum antiferromagnet. vol 8, pp. 15148
13. Albaladejo M, Oller JA (2012) Size of the σ meson and its nature. Phys Rev D 86:034003
14. Cea T, Benfatto L (2014) Nature and raman signatures of the Higgs amplitude mode in the coexisting superconducting and charge-density-wave state. Phys Rev B 90:224515
15. Cea T, Castellani C, Seibold G, Benfatto L (2015) Nonrelativistic dynamics of the amplitude (Higgs) mode in superconductors. Phys Rev Lett 115:157002
16. Krull H, Bittner N, Uhrig GS, Manske D, Schnyder AP (2016) Coupling of Higgs and Leggett modes in non-equilibrium superconductors. Nat Commun 7:11921
17. Barger V, Ishida M, Keung W (2012) Total width of 125 GeV Higgs boson. Phys Rev Lett 108:261801
18. Halperin WP (1982) Acoustic order parametermode spectroscopy in superfluid 3he-b. Physica B+C, 109:1596–1605
19. Bayrakci SP, Keller T, Habicht K, Keimer B (2006) Spin-wave lifetimes throughout the brillouin zone. Science 312(5782):1926–1929
20. Kuroe H, Kusakabe K, Oosawa A, Sekine T, Fumiko Y, Tanaka H, Matsumoto M (2008) Magnetic field induced one-magnon Raman scattering in the magnon Bose-Einstein condensation phase of TlCuCl3. Phys Rev B 77:134420
21. Kuroe H, Oosawa A, Sekine T, Yamada F, Tanaka H, Matsumoto M (2009) Anticrossing effect between magnon and phonon in Bose-Einstein condensation phase of $TlCuCl_3$. J Phys Conf Ser 150(4):042104
22. We note that Ref. [21] fits a linewidth to the Raman spectra of the Higgs mode. However, it is not specified whether this represents the intrinsic decay width of the Higgs mode, or if it represents an effective linewidth including instrument resolution
23. Ch Ruegg N, Cavadini A, Furrer H-U, Gudel K, Kramer H, Mutka A, Wildes KH, Vorderwisch P (2003) Bose-Einstein condensation of the triplet states in the magnetic insulator $TlCuCl_3$. Nature 423(6935):62–65
24. Jaime M, Correa VF, Harrison N, Batista CD, Kawashima N, Kazuma Y, Jorge GA, Stern R, Heinmaa I, Zvyagin SA, Sasago Y, Uchinokura K (2004) Magnetic-field-induced condensation of triplons in Han purple pigment $BaCuSi_2O_6$. Phys Rev Lett 93:087203
25. Kofu M, Kim J-H, Ji S, Lee S-H, Ueda H, Qiu Y, Kang H-J, Green MA, Ueda Y (2009) Weakly coupled S = 1/2 quantum spin singlets in $Ba_3Cr_2O_8$. Phys Rev Lett 102:037206

26. Wang Z, Quintero-Castro DL, Zherlitsyn S, Yasin S, Skourski Y, Islam ATMN, Lake B, Deisenhofer J, Loidl A (2016) Field-induced magnonic liquid in the 3D spin-dimerized antiferromagnet $Sr_3Cr_2O_8$. Phys Rev Lett 116:147201
27. Okada M, Tanaka H, Kurita N, Johmoto K, Uekusa H, Miyake A, Tokunaga M, Nishimoto S, Nakamura M, Jaime M, Radtke G, Saúl A (2016) Quasi-two-dimensional Bose-Einstein condensation of spin triplets in the dimerized quantum magnet $Ba_2CuSi_2O_6Cl_2$. Phys Rev B 94:094421
28. Grundmann H, Sabitova A, Schilling A, von Rohr F, Frster T, Peters L (2016) Tuning the critical magnetic field of the triplon Bose-Einstein condensation in $Ba_{3-x}Sr_xCr_2O_8$. New J Phys 18(3):033001
29. Kurita N, Tanaka H (2016) Magnetic-field- and pressure-induced quantum phase transition in $CsFeCl_3$ proved via magnetization measurements. Phys Rev B 94:104409
30. Brambleby J, Goddard PA, Singleton J, Jaime M, Lancaster T, Huang L, Wosnitza J, Topping CV, Carreiro KE, Tran HE, Manson ZE, Manson JL (2017) Adiabatic physics of an exchange-coupled spin-dimer system: magnetocaloric effect, zero-point fluctuations, and possible two-dimensional universal behavior. Phys Rev B 95:024404
31. Giamarchi T, Tsvelik AM (1999) Coupled ladders in a magnetic field. Phys Rev B 59:11398–11407
32. Nikuni T, Oshikawa M, Oosawa A, Tanaka H (2000) Bose-Einstein condensation of dilute magnons in $TlCuCl_3$. Phys Rev Lett 84:5868–5871
33. Matsumoto M, Normand B, Rice TM, Sigrist M (2002) Magnon dispersion in the field-induced magnetically ordered phase of $TlCuCl_3$. Phys Rev Lett 89:077203
34. Matsumoto M, Normand B, Rice TM, Sigrist M (2004) Field- and pressure-induced magnetic quantum phase transitions in $TlCuCl_3$. Phys Rev B 69:054423
35. Utesov OI, Syromyatnikov AV (2014) Theory of field-induced quantum phase transition in spin dimer system $Ba_3Cr_2O_8$. J Magn Magn Mater 358359:177–182
36. Scammell HD, Sushkov OP (2017) Multiple universalities in order-disorder magnetic phase transitions. Phys Rev B 95:094410
37. Giamarchi T, Ruegg C, Tchernyshyov O (2008) Bose-Einstein condensation inmagnetic insulators. Nat Phys 4(3):198–204
38. Zapf V, Jaime M, Batista CD (2014) Bose-Einstein condensation in quantum magnets. Rev Mod Phys 86:563–614
39. Qin YQ, Normand B, Sandvik AW, Meng ZY (2017) Amplitude mode in three-dimensional dimerized antiferromagnets. Phys Rev Lett 118:147207
40. Lohöfer M, Wessel S (2017) Excitation-gap scaling near quantum critical three-dimensional antiferromagnets. Phys Rev Lett 118:147206
41. Considering the AFM part of the phase diagram Fig. 8.1(b) one has to remember that an arbitrary weak magnetic field orients the AFM macroscopic polarization perpendicular to the field
42. Rüegg C, Normand B, Matsumoto M, Furrer A, McMorrow DF, Krämer KW, Güdel HU, Gvasaliya SN, Mutka H, Boehm M (2008) Quantum magnets under pressure: controlling elementary excitations in $TlCuCl_3$. Phys Rev Lett 100:205701
43. Rüegg C, Cavadini N, Furrer A, Krämer K, Güdel HU, Vorderwisch P, Mutka H (2002) Spin dynamics in the high-field phase of quantum-critical S=1/2 $TlCuCl_3$. Appl Phys A 74(1):s840–s842
44. Fisher DS (1989) Universality, low-temperature properties, and finite-size scaling in quantum antiferromagnets. Phys Rev B 39:11783–11792
45. Sachdev S (2011) Quantum phase transitions. Cambridge University Press
46. Affleck I (1991) Bose condensation in quasi-one-dimensional antiferromagnets in strong fields. Phys Rev B 43:3215–3222
47. To plot the B-dependence we use $g = 2.2$ to fit the data. The value of g follows from the critical field $B_c = 5.7$T and the gap $\Delta = m_0 = 0.71$meV measured in Ref. [43]. Note that Ref. [50] presents a different value of the critical field, $B_c = 4.73$T. So, B_c might be sample dependent

48. Farutin AM, Marchenko VI (2007) Dynamics of paramagnets at zero temperature. J Exp Theor Phys 104(5):751–757
49. Kulik Y, Sushkov OP (2011) Width of the longitudinal magnon in the vicinity of the O(3) quantum critical point. Phys Rev B 84:134418

Chapter 9
Bose-Einstein Condensation of Particles with Half-Integer Spin

Abstract We consider the magnetic field induced Bose-condensation of bosonic particles with spin 1/2. We derive properties of the condensate in both two and three dimensions, focusing on mode dispersions, phase boundaries, and critical indices. We find that the unusual condensate supports a Goldstone mode with quadratic dispersion, and this provides a "smoking gun" criterion for searches of the novel phase of matter.

9.1 Introduction

The famous spin-statistics theorem [1] claims that particles with integer spin obey Bose-statistics while particles with half-integer spin obey Fermi-statistics. Therefore, for all particles restricted to obey the spin-statistics theorem only those particles with integer spin can undergo Bose-condensation. However, as introduced in Chap. 1, certain condensed matter systems, e.g. some kinds of spin liquids [2] as well as the deconfined quantum criticality (DQC) scenario [3, 4], are predicted to host bosonic quasiparticles with spin 1/2.

In this chapter, we assume the DQC theoretical framework and consider the application of an external magnetic field to drive the system to an exotic Bose-condensate (BEC) of spin 1/2 DQC spinons. This work represents the first suggestion of such an exotic condensate, and as such our primary motivation will be to derive the key properties/observables that would allow the exotic phase to be identified in future experimental or numerical studies.

Before proceeding, it is important to state that bosonic spinons do not contradict the spin-statistics theorem. First, they are quasiparticles in a solid and hence the Lorentz invariance is explicitly violated. Second, they are quasiparticles with a nonperturbative origin; only quasiparticles of a nonperturbative origin can violate the spin statistics theorem. Quasiparticles that can be adiabatically (perturbatively) transferred from solid/medium to vacuum obey standard statistics. Electrons are always fermionic. Phonons (atomic displacements) and magnons (spin deflections) are always bosonic.

H. Scammell, *Interplay of Quantum and Statistical Fluctuations in Critical Quantum Matter*, Springer Theses,
https://doi.org/10.1007/978-3-319-97532-0_9

As discussed in detail in the introduction, Sect. 1.3, there is mounting numerical support for the DQC scenario—we will not repeat the discussion here. Instead we will just state the essential numerical results obtained on the JQ model, which, to the best of our knowledge, enables the analysis of the largest lattices among all existing DQC and spin liquid lattice models.

The DQC scenario was directly addressed in a QMC study Ref. [5]. An analysis of the results through the lens of DQC quantum field theory demonstrated the following: (i) the spin of the bosonic quasiparticle is really $S = 1/2$. This is a major conclusion. (ii) The spinons are weakly interacting. (iii) The emergent $U(1)$ gauge field proposed in [3, 4] does not show up as a dynamic variable. (iv) There is a logarithmic correction to one of the observables (the Wilson ratio). The correction is due to large spinon occupation numbers in a weak magnetic field, $n_k \gg 1$, and hence, the correction is a precursor to spinon Bose-condensation.

Armed with this information, we can now proceed to an effective quantum field theory description of the DQC spinons in an external magnetic field.

9.2 Theory of the Spinon BEC

Following Refs. [3, 4] we use the CP^1 representation to describe spinons, so the mathematical object of the theory, z, is an $SU(2)$ spinor. The effective Lagrangian of spinons is

$$\mathscr{L} = \{\partial_t z^\dagger + iSz^\dagger(\vec{\sigma}\cdot\vec{B})\}\{\partial_t z - iS(\vec{\sigma}\cdot\vec{B})z\} - (\nabla z^\dagger)(\nabla z) - m^2 z^\dagger z - \frac{\alpha}{2}(z^\dagger z)^2. \tag{9.1}$$

We set the spinon speed equal to unity, as well as $gS\mu_B = 1$, where g is the gyromagnetic factor, $S = 1/2$ is spin of the spinon, and μ_B is the Bohr magneton. The $\vec{B}$ is the external uniform static magnetic field, $\vec{\sigma}$ are the usual Pauli matrices, and m is mass of the spinon. The interaction is repulsive, $\alpha > 0$, and first we assume that $m^2 \geq 0$. Unlike Refs. [3, 4] the Lagrangian (9.1) does not contain a dynamic gauge field. This is directly motivated by the QMC analysis [5] which does not show a contribution of the gauge field. While the DQC motivation comes from two-dimensional (2D) systems (spatial dimensions), the Lagrangian (9.1) can be considered both in two and three dimensions.

Consider the case when $B < m$: the classical expectation of the field z is zero, and the system is disordered. We are interested in the quantum fluctuations in this phase, for which we solve the linearised Euler-Lagrange equation

$$\ddot{z} - \nabla^2 z - 2i(\vec{\sigma}\cdot\vec{B})\dot{z} + \left(m^2 - B^2\right)z = 0 \tag{9.2}$$

By ignoring the interaction term, we have in mind that it is reabsorbed in the quantum renormalisation of the mass term $m \to m_\Lambda$. The plane wave solution is $z(t, \boldsymbol{x}) \to u_\lambda e^{i\boldsymbol{k}\cdot\boldsymbol{x} - i\omega_{k,\lambda}t}$, where the spinor u_λ follows from $(\vec{\sigma}\cdot\vec{B})u_\lambda = \lambda B u_\lambda$,

$\lambda = \pm 1$. Hence,

$$\omega_{\boldsymbol{k},\lambda} = \Omega_{\boldsymbol{k}} + \lambda B \; . \tag{9.3}$$

Here $\Omega_{\boldsymbol{k}} = \sqrt{m^2 + \boldsymbol{k}^2}$. The standard canonical quantisation gives,

$$z = \sum_{\boldsymbol{k},\lambda} \frac{1}{\sqrt{2\Omega_{\boldsymbol{k}}}} u_\lambda \left[\hat{a}_{\boldsymbol{k},\lambda} e^{i\omega_{\boldsymbol{k},\lambda}t - i\boldsymbol{k}\cdot\boldsymbol{x}} + \hat{b}^\dagger_{\boldsymbol{k},\lambda} e^{-i\omega_{\boldsymbol{k},\lambda}t + i\boldsymbol{k}\cdot\boldsymbol{x}} \right] \tag{9.4}$$

where, $\hat{a}_{\boldsymbol{k},\lambda}$ and $\hat{b}_{\boldsymbol{k},\lambda}$ are bosonic creation and annihilation operators. Importantly, z is a usual spinor, $z \neq z^\dagger$, and therefore one must introduce two types of bosons, $\hat{a}$ and $\hat{b}$, or spinons and antispinons. This implies that at a given momentum there are four degrees of freedom due to the combinations $(\hat{a}, \hat{b}) \times (\lambda = \pm 1)$ [3, 4].

Consider the case when $B > m$: the system undergoes Bose-condensation (the dispersion (9.3) becomes negative at $\boldsymbol{k} = 0$). The energy density corresponding to (9.1) is $E = \dot{z}^\dagger \dot{z} + \nabla z^\dagger \nabla z + (m^2 - B^2) z^\dagger z + \frac{\alpha}{2}(z^\dagger z)^2$. Minimisation of the energy gives the spin condensate and the classical energy, see also Appendix G

$$z_0^\dagger z_0 = \frac{(B^2 - m^2)}{\alpha} \; , \quad E_0 = -\frac{(B^2 - m^2)^2}{\alpha} . \tag{9.5}$$

This expression has the following implications. First of all, $z_0^\dagger z_0$ is a real number that effectively counts together the number of spinons and antispinons. However, the condensate $z_0^\dagger z_0$ gives no information as to the relative contributions of spinons and antispinons. The second implication is that the ground state energy has no dependence on how the spin polarisation vector $\vec{\zeta}_0 = z_0^\dagger \vec{\sigma} z_0 = (\sin\theta_0 \cos\varphi_0, \sin\theta_0 \sin\varphi_0, \cos\theta_0)$ is directed relative to the magnetic field ($\vec{B} || \hat{z}$), see Fig. 9.1. To obtain this expression, we have represented the spin condensate as,

$$z_0 = A_0 e^{i\gamma_0} \begin{pmatrix} \cos\theta_0/2 \\ e^{i\varphi_0} \sin\theta_0/2 \end{pmatrix} , \tag{9.6}$$

where $A_0 = \sqrt{(B^2 - m^2)/\alpha}$.

Despite the degeneracy of ground state energy on the orientation $\vec{\zeta}_0$, we are happy to find that the induced magnetisation is indeed directed along $\vec{B}$, $\vec{M} = -\frac{\partial E_0}{\partial B} = 4\frac{(B^2 - m^2)}{\alpha} \vec{B}$. The degeneracy of the spin alignment with respect to the magnetic field is a consequence of the system containing both particles and antiparticles. But more importantly, the degeneracy of $\vec{\zeta}_0$ is intimately linked to the gapless Goldstone excitations, which correspond to variations of the angles θ_0 and φ_0 in Fig. 9.1. We will see below via explicit calculation that there are indeed two Goldstone modes with precisely these properties. As a final comment on the degeneracy, we note that if an emergent $U(1)$ gauge field were included in the Lagrangian (9.1) then, via the Higgs mechanism, one of the Goldstone modes would vanish from the spectrum

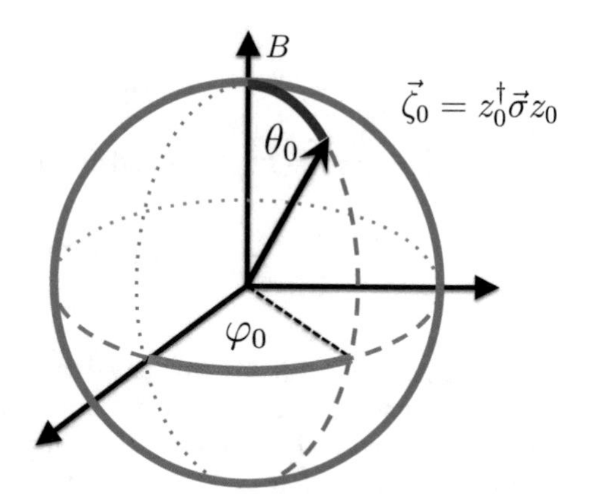

Fig. 9.1 Orientation of the spin condensate vector $\vec{\zeta}_0$ with respect to magnetic field

implying that the degeneracy with respect to variations of the angles θ_0 and φ_0 would be (partially) lifted. We leave a detailed analysis of such a scenario for future work.

We now wish to find the excitations of the novel spinon condensate phase. Let us denote fluctuations about the condensate (9.6) as $z = z_0 + \delta z$. The Euler-Lagrange equation for δz reads

$$\delta\ddot{z} - \nabla^2 \delta z - 2i(\vec{\sigma} \cdot \vec{B})\delta\dot{z} + \alpha(\delta z^\dagger z_0) z_0 + \alpha(z_0^\dagger \delta z) z_0 = 0 \tag{9.7}$$

The solution takes the form, $\delta z = z_+ e^{i\omega t - ikr} + z_- e^{-i\omega t + ikr}$. After some algebra, presented in Appendix G, we find the following modes

$$\begin{aligned} \omega_{1k} &= \sqrt{B^2 + \boldsymbol{k}^2} - B \\ \omega_{2k} &= \sqrt{3B^2 - m^2 + \boldsymbol{k}^2 - \sqrt{(3B^2 - m^2)^2 + 4B^2\boldsymbol{k}^2}} \\ \omega_{3k} &= \sqrt{B^2 + \boldsymbol{k}^2} + B \\ \omega_{4k} &= \sqrt{3B^2 - m^2 + \boldsymbol{k}^2 + \sqrt{(3B^2 - m^2)^2 + 4B^2\boldsymbol{k}^2}}\,. \end{aligned} \tag{9.8}$$

The first two modes are gapless (Goldstone), $\omega_{\boldsymbol{k}=0} = 0$, while the second two modes are gapped. We call the modes corresponding to ω_{3k} and ω_{4k}, the precession and the Higgs modes. Surprisingly the first Goldstone mode has quadratic dispersion at small $\boldsymbol{k}$, $\omega_{1k} \approx \frac{\boldsymbol{k}^2}{2B}$, while the second Goldstone mode has the usual linear dispersion, $\omega_{2k} \approx ck$, $c = \sqrt{\frac{B^2 - m^2}{3B^2 - m^2}}$.

In order to gain a visual understanding of the excitation modes, we appeal to the spin vector representation and look at the variations it receives due to each mode, i.e. $\vec{\zeta} = z^\dagger \vec{\sigma} z = \vec{\zeta}_0 + \delta\vec{\zeta}$ where $\delta\vec{\zeta} \approx z_0^\dagger \vec{\sigma} \delta z + \delta z^\dagger \vec{\sigma} z_0$. Calculations presented in Appendix G show that for the Goldstone modes at small momenta $\boldsymbol{k} \to 0$, the variations are

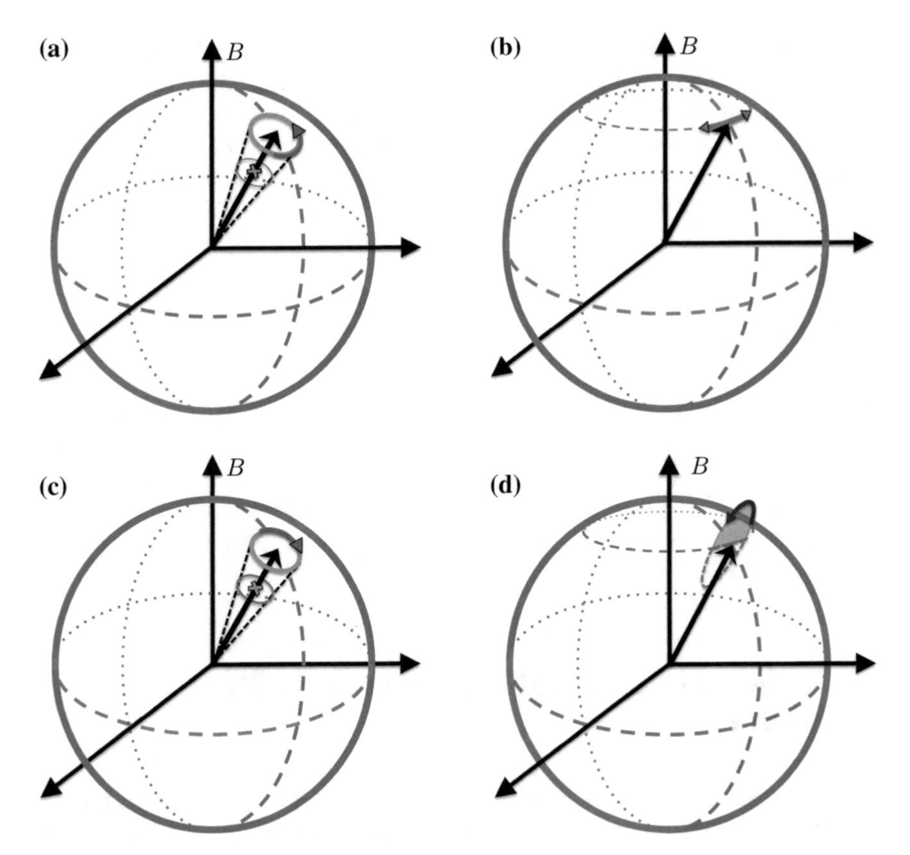

Fig. 9.2 Oscillations of the spin polarisation vector $\vec{\zeta}$ with respect to the condensate vector $\vec{\zeta}_0$ in different excitation modes. The solid black arrow represents an arbitrary $\vec{\zeta}_0$. **a** Goldstone mode with quadratic dispersion. Anticlockwise rotation, the orange circle shows the path traced by the oscillations. **b** Goldstone mode with linear dispersion. Linear oscillations in the direction perpendicular to the meridian. **c** Precession mode $|3\rangle$. Clockwise rotation, the orange circle shows the path traced by the oscillations. **d** Higgs mode $|4\rangle$. Elliptic oscillations in the plane perpendicular the meridian

$$\delta\vec{\zeta}_1 \propto \begin{pmatrix} \cos\theta_0 \cos(\psi + \varphi_0) \\ \cos\theta_0 \sin(\psi + \varphi_0) \\ -\sin\theta_0 \cos\psi \end{pmatrix}, \qquad \delta\vec{\zeta}_2 \propto \begin{pmatrix} -\sin\theta_0 \sin\varphi_0 \\ \sin\theta_0 \cos\varphi_0 \\ 0 \end{pmatrix} \sin\psi. \tag{9.9}$$

Here ψ is the plane wave phase, $\psi = \omega t - \boldsymbol{k} \cdot \boldsymbol{x}$. For the quadratic Goldstone mode, the variation $\delta\vec{\zeta}_1$ is shown in Fig. 9.2a. It represents an anticlockwise rotation of vector $\vec{\zeta}$ around the vacuum polarisation vector $\vec{\zeta}_0$. The spin polarisation variation $\delta\vec{\zeta}_2$ in the linear Goldstone mode is shown in Fig. 9.2b. The mode represents linear oscillations of vector $\vec{\zeta}$ around the vacuum polarisation vector $\vec{\zeta}_0$ in the direction perpendicular to the meridian.

Both Goldstone modes satisfy the orthogonality condition $\delta\vec{\zeta}_1 \cdot \vec{\zeta}_0 = \delta\vec{\zeta}_2 \cdot \vec{\zeta}_0 = 0$. The spin oscillations are in the plane orthogonal to the spin condensate vector, and hence correspond to variations of the angles θ_0 and φ_0—as anticipated. Polarisations of the Goldstone modes are intuitive; circular polarisation for the quadratic mode, similar to a ferromagnet, and linear polarisation in the linear mode, similar to an antiferromagnet. An interesting consequence of the quadratic mode is that condensate does not support superfluidity. The Landau criterion of superfluidity is not fulfilled.

Now let us consider the gapped precession and Higgs modes. The corresponding variations of the spin expectation vector are found to be, see Appendix G for details,

$$\delta\vec{\zeta}_3 \propto \begin{pmatrix} -\cos\theta_0\cos(\psi-\varphi_0) \\ \cos\theta_0\sin(\psi-\varphi_0) \\ \sin\theta_0\cos\psi \end{pmatrix}, \qquad \delta\vec{\zeta}_4 \propto \begin{pmatrix} -\sin\theta_0\sin\varphi_0 \\ \sin\theta_0\cos\varphi_0 \\ 0 \end{pmatrix} \sin\psi + b_4\vec{\zeta}_0\cos\psi, \tag{9.10}$$

where b_4, presented in Appendix G, is a non-zero constant $b_4 \to const \neq 0$ at $\mathbf{k} \to 0$. The third mode represents a clockwise rotation of vector $\vec{\zeta}$ around the vacuum polarisation vector $\vec{\zeta}_0$, Fig. 9.2c, it is truly a *precessing* mode. Finally, the fourth mode represents elliptic oscillations in the plane perpendicular to the meridian, Fig. 9.2d. This mode therefore possesses a longitudinal component (parallel to $\vec{\zeta}_0$), and the terminology *Higgs* is justified.

9.3 Renormalization, Critical Indices and the Phase Diagram

We now consider the case where $B = 0$ and $m^2 < 0$. This can be achieved (at least in principle) by taking $m^2 \propto p_c - p$ in the Lagrangian (9.1), such that p is the quantum tuning parameter and $p = p_c$ is the quantum critical point. For this QPT straightforward algebra gives three Goldstone excitations and one gapped Higgs excitation

$$\omega_1 = \omega_2 = \omega_3 = k$$
$$\omega_4 = \sqrt{2|m^2| + k^2}\,.$$

We will call this phase the *spinon antiferromagnet*. With this result in mind, we can now look at the generic properties of the (p, B, T) phase diagram of spinons.

9.3.1 Phase Diagram

Specialising our discussion to 3+1 dimensions, we will now discuss and uncover previously unknown properties of the spinon phase diagram, i.e. we will categorise the universal features of this exotic magnetic condensate. We are interested in the interplay between the three tuning handles of the phase diagram: the quantum tuning

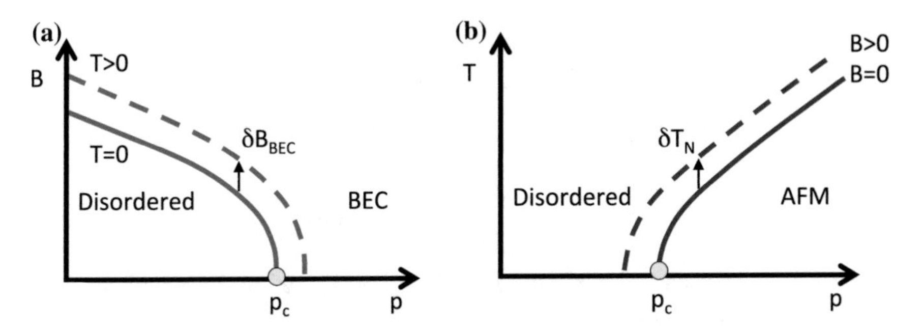

Fig. 9.3 Critical field and temperature power law shifts. **a** Shift of critical field-pressure line with temperature $\delta B_{BEC} \sim T^{\phi}$. Solid blue curve is at zero temperature, dashed blue at non-zero temperature. **b** Shift of critical (Néel) temperature-pressure line with field $\delta T_N \sim B^{1/\phi}$. Solid red curve is at zero field, dashed red at non-zero field

$p_c - p$, the magnetic field, B, and the temperature. In Fig. 9.3 we present the generic phase diagrams of the spinons either condensing via tuning B to form a BEC, or by tuning p to generate a spinon AFM-like phase. Panel (a) shows the spinon Bose condensation (BEC) line in the field-pressure diagram, and panel (b) shows the spinon antiferromagnetic (AFM) transition line in the temperature-pressure diagram. As in Chap. 7, the point of primary interest is the critical field-critical temperature power law,

$$\mathrm{a} : \delta B_{BEC} \sim T^{\phi}, \qquad\qquad \mathrm{b} : \delta T_N \sim B^{1/\phi}, \tag{9.11}$$

The shift of the BEC transition line at small temperature is shown schematically in Fig. 9.3a; while the shift of the AFM/Néel transition line at small field is in Fig. 9.3b. We will now calculate the evolution of the critical index ϕ.

9.3.2 Renormalization

The condition for the critical curves in the phase diagram, e.g. Fig. 9.3, is given by

$$m_{\Lambda}^2(T_c) - B_c^2 = 0 \tag{9.12}$$

where the one-loop renormalised mass is found to be

$$m_{\Lambda,\pm}^2 = m_0^2 \left[\frac{\alpha_\Lambda}{\alpha_0}\right]^{\frac{N+2}{N+8}} + \frac{1}{2}(N+2)\alpha_\Lambda \Sigma_T$$

$$\Sigma_T \equiv \sum_k 1/\omega_k^0 \{n(\omega_k^+) + n(\omega_k^-)\}, \tag{9.13}$$

$$\alpha_\Lambda = \frac{\alpha_0}{1 + (N+8)\alpha_0/(16\pi^2)\ln(\Lambda_0/\Lambda)}. \tag{9.14}$$

Here $N = 4$ (discussed next), $n(\omega_{\boldsymbol{k}}) = 1/(e^{\frac{\omega_{\boldsymbol{k}}}{T}} - 1)$, and we introduce the function Σ_T to denote the nonzero temperature part of the self-energy, see related discussions in Chap. 7. The entire renormalization group procedure mimics that developed in Chap. 6. To most easily see this, one can perform the mapping

$$z = \frac{1}{\sqrt{2}} \begin{pmatrix} \varphi_1 + i\varphi_2 \\ \varphi_3 + i\varphi_4 \end{pmatrix} \tag{9.15}$$

to obtain the following Lagrangian

$$\mathscr{L} = \frac{1}{2}\vec{\varphi}^T \begin{pmatrix} [\omega^2 - 2i\omega B - k^2 - (m^2 - B^2)]\,\mathbb{I}_2 & 0 \\ 0 & [\omega^2 + 2i\omega B - k^2 - (m^2 - B^2)]\,\mathbb{I}_2 \end{pmatrix} \vec{\varphi} - \frac{\alpha}{8}\vec{\varphi}^4 \tag{9.16}$$

where $\vec{\varphi}^T = (\varphi_1, \varphi_2, \varphi_3, \varphi_4)$ are real valued fields and $\mathbb{I}_2$ is the 2×2 identity matrix. We see that this Lagrangian has a similar structure to that seen in Chaps. 6 and 7, except now the number of real scalar fields is $N = 4$. The renormalization runs analogously to that presented in those chapters, however in the present case we see that there is an added convenience—the matrix kernel (inverse Greens function) is already diagonal in the cartesian representation.

9.3.3 Critical Indices

There are three distinct cases: (I) Above the critical pressure, when $T_c = T_N$ i.e. critical temperature equals the AFM/Néel temperature; (II) exactly at the critical pressure, $p = p_c$; (III) below the critical pressure, when $T_c = T_{BEC}$.

Consider case (I); $p > p_c$. In this case according to Eq. (9.11b) the Néel temperature varies in a weak magnetic field. To calculate Σ_T at $B \to 0$ we take the critical line dispersions $\omega_{\boldsymbol{k}}^+ = \omega_{\boldsymbol{k}}^- = \omega_{\boldsymbol{k}}^0 = k$. Hence $\Sigma_T = \frac{1}{12}T^2$, where $T = T_{N0} + \delta T_N$; T_{N0} is the Néel temperature in zero magnetic field. Hence using Eq. (8.7) we find

$$\text{(I)}: \delta T_N = \frac{12}{(N+2)\alpha_\Lambda} \frac{B^2}{T_{N0}} \qquad \text{at } B \ll T_{N0}. \tag{9.17}$$

So the critical index in Eq. (9.11b) is $\phi = 1/2$.

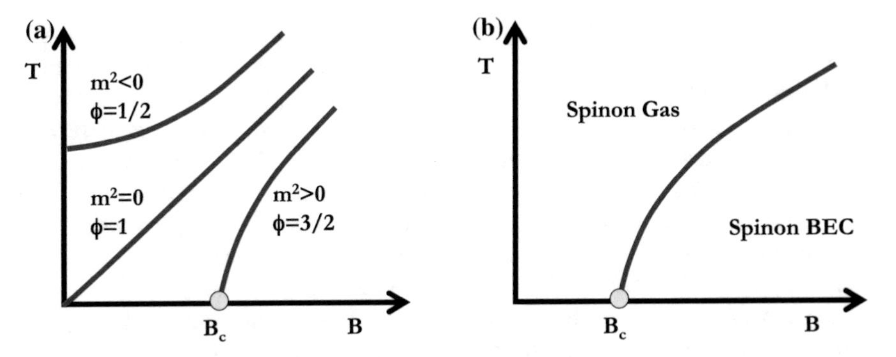

Fig. 9.4 Multiple universalities: Various curves show the critical field $B_c(T)$ for three distinct cases: $m^2 < 0$, $m^2 = 0$, and $m^2 > 0$. The curves are purely schematic

Consider case (II); tuning exactly to the quantum critical point, $p = p_c$, $T_{N0} = 0$. Again, to calculate Σ_T at $B \to 0$ we have to take the critical line dispersions $\omega_{\boldsymbol{k}}^+ = \omega_{\boldsymbol{k}}^- = \omega_{\boldsymbol{k}}^0 = k$ and hence again $\Sigma_T = \frac{1}{12}T^2$. Substitution into Eq. (8.7) gives

$$\text{(II)}: B_c = \sqrt{\frac{(N+2)\alpha_\Lambda}{24}}T \qquad \text{at } B_c \ll T. \tag{9.18}$$

The condition $B_c \ll T$ is satisfied at sufficiently low temperatures since the coupling constants decays logarithmically, $\alpha_\Lambda \propto 1/\ln\left(\frac{\Lambda_0}{T}\right)$. Hence in this case (II), the critical index of Eq. (9.11) is $\phi = 1$, and we find that, in addition to the exponent, there is nontrivial logarithmic scaling. In Fig. 9.4 we illustrate the asymptotic (9.18) by straight line labelled $\phi = 1$, which originates from $B = T = 0$.

Finally we consider the BEC case (III), $p < p_c$. In this case only the $\omega_{\boldsymbol{k}}^-$ dispersion branch is critical, $\omega_{\boldsymbol{k}}^- \approx \frac{k^2}{2\Delta_0}$, where Δ_0 is the gap at $B = 0$. The other modes are gapped. Calculation of Σ_T gives $\Sigma_T = \alpha_\Lambda \frac{\zeta(3/2)}{\pi\sqrt{2\pi}}\sqrt{\Delta_0}T^{3/2}$, where ζ is Riemann's ζ-function. Hence, using Eq. (8.7) we find

$$\text{(III)}: \frac{\delta B_c}{\Delta_0} = \alpha_\Lambda \frac{(N+2)\zeta(3/2)}{10(2\pi)^{\frac{3}{2}}}\left(\frac{T}{\Delta_0}\right)^{3/2} \quad \text{at } \delta B_c \ll \Delta_0. \tag{9.19}$$

As expected the critical index in Eq. (9.11a) is $\phi = 3/2$.

Regimes (I) and (II) have only been considered before by the present author in Chap. 7. It is interesting that the same universality carries over to the spinon theory. Note, we have not attempted to plot the result derived here explicitly, like we did in Chap. 7, since unlike there we do not have fitted values for the coupling constant and normalisation point (α_0, Λ_0). Instead, it suffices to plot a schematic phase diagram—depicting the three cases discussed here.

The existence of three critical exponents $\phi = 3/2$, 1 and 1/2, and even logarithmic corrections to these exponents, is certainly a testable result. However, since these critical exponents are shared by the analogous magnon (spin $S = 1$) theory, it is not a conclusive test of the exotic spinon BEC scenario. Instead, in the next subsection we discuss a much simpler and direct (smoking gun) test of the spinon BEC.

More importantly, we note that when applying the present analysis to the AFM $\rightarrow$ VBS transition in a quantum magnet one must remember that the deconfined description is valid only within a vicinity of the quantum critical point. There is no doubt that deep inside the VBS or AFM phase the quasiparticles are usual triplons or magnons, respectively. This implies that there is a spinon confinement length ξ which depends on the detuning from the quantum critical point. It is possible that the gauge field [3, 4], despite not appearing as a dynamic variable in (9.1), still acts as a constraint contributing to formation of the confinement length. The deconfined description, and in particular Eq. (9.20), is valid if the spinon thermal wavelength is smaller than the confinement scale, $\lambda_T \sim 1/\sqrt{BT} \ll \xi$.

9.3.4 *The Smoking Gun*

Finally, we point out that the appearance of a Goldstone with quadratic dispersion is one of the most intriguing properties of the spinon Bose-Condensate. In a conventional Bose-condensate, say of magnons, the specific heat scales quadratically with temperature, $C \propto T^2$, in two spatial dimensions and scales cubically with temperature, $C \propto T^3$ in three spatial dimensions. On the other hand, for the spinon Bose-condensate

$$
\begin{aligned}
2D: \quad C &= \frac{1}{\pi}\zeta(2)SBT \\
3D: \quad C &= \frac{15}{8\sqrt{2}\pi^{3/2}}\zeta(5/2)(SB)^{3/2}T^{3/2} \, .
\end{aligned}
\tag{9.20}
$$

Here $\zeta(x)$ is Riemann's zeta function, and we have explicitly reinstated $S = 1/2$, which was absorbed into the definition of B in all previous equations. We therefore claim that the unconventional temperature scaling of the specific heat is a smoking gun for the spinon Bose-condensate. Note that in our expression for the 2D case (9.20), we assume a condensate still exists at finite temperature. Of course we need to keep in mind the Mermin-Wagner theorem [6], which excludes the existence of true long range order at $T \neq 0$ in 2D case. Instead the *quasi*-condensate has exponentially large correlation length instead of infinite range, and ultimately this does not influence the power of temperature in the specific heat (9.20).

9.4 Conclusion

We have proposed an exciting scenario whereby particles with spin $S = 1/2$ form a Bose-condensate. We analyse the condensate in both two and three spatial dimensions and find the key observables, such as dispersions of all modes in the condensate and disordered phases, the phase boundaries and critical indices, and some properties of the condensate itself. The most intriguing finding is a quadratic Goldstone mode in the condensate phase. We argue that this implies unconventional temperature scaling of the specific heat, and hence may be used as a diagnostic of this novel phase of matter.

References

1. Pauli W (1940) The connection between spin and statistics. Phys Rev 58:716–722
2. Wen X-G (2002) Quantum orders and symmetric spin liquids. Phys Rev B 65:165113
3. Senthil T, Vishwanath A, Balents L, Sachdev S, Fisher MPA (2004) Deconfined quantum critical points. Science 303(5663):1490–1494
4. Senthil T, Balents L,Sachdev S, Vishwanath A, Fisher MPA (2004) Quantum criticality beyond the Landau-Ginzburg-Wilson paradigm. Phys Rev B 70:144407
5. Sandvik AW, Kotov VN, Sushkov OP (2011) Thermodynamics of a gas of deconfined bosonic spinons in two dimensions. Phys Rev Lett 106:207203
6. Mermin ND, Wagner H (1966) Absence of ferromagnetism or antiferromagnetism in one- or two-dimensional isotropic Heisenberg models. Phys Rev Lett 17:1133–1136

Appendix A
Appendix for Chapter 2

In this appendix we discuss the derivation of renormalised mass and running coupling constant, as well as the Néel temperature, and the influence of spin-orbit anisotropy.

A.1 Running Coupling Constant

The four point vertex in Fig. (2.2) is calculated to second order in α (with a Euclidean metric)

$$\begin{aligned}\Gamma^{(4)} &= \alpha - (N+8)\alpha^2 \int_{\Lambda}^{\Lambda_c} \frac{d^4k}{(2\pi)^4} \frac{1}{k^4} \\ &= \alpha - \frac{(N+8)\alpha^2}{8\pi^2} \ln\left(\frac{\Lambda_c}{\Lambda}\right).\end{aligned} \tag{A.1}$$

The infrared cut-off, Λ, is given by the mass gap, or the temperature scale. We use a Callan-Symanzik equation to find the Beta function

$$\begin{aligned}\left[\frac{d}{d\ln(\Lambda_c/\Lambda)} + \beta(\alpha)\frac{d}{d\alpha}\right]\Gamma^{(4)} &= 0 \\ \Rightarrow \beta(\alpha) &= \frac{(N+8)\alpha^2}{8\pi^2} \\ \Rightarrow \frac{d\alpha}{d\ln(\Lambda_0/\Lambda)} &= -\frac{(N+8)\alpha^2}{8\pi^2} \\ \alpha_\Lambda &= \frac{\alpha_0}{1 + \frac{(N+8)\alpha_0}{8\pi^2}\ln(\Lambda_0/\Lambda)}\end{aligned} \tag{A.2}$$

where Λ_c is some momentum cut-off such as the inverse lattice spacing, while Λ_0 is the normalisation point.

H. Scammell, *Interplay of Quantum and Statistical Fluctuations in Critical Quantum Matter*, Springer Theses,
https://doi.org/10.1007/978-3-319-97532-0

A.2 Self-energy in the Disordered Phase

Approaching from the disordered phase, the first perturbative correction to the triplon gap comes from the one loop self-energy

$$\begin{aligned}\Sigma(\Delta, T) &= (N+2)\alpha_\Lambda \sum_{\mathbf{k}} \frac{1}{\omega_k}\left[\frac{1}{2} + \frac{1}{e^{\frac{\omega_k}{T}} - 1}\right] \\ &= (N+2)\alpha_\Lambda \int \frac{d^3k}{(2\pi)^3}\frac{1}{2\omega_k} \\ &+ (N+2)\alpha_\Lambda \int \frac{d^3k}{(2\pi)^3}\frac{1}{\omega_k}\frac{1}{(e^{\frac{\omega_k}{T}} - 1)}.\end{aligned} \tag{A.3}$$

The coupling constant coefficient is the running coupling α_Λ, since the two point corrections are multiplicative with the four point vertices. With these corrections the triplon gap becomes dependent on both p and T

$$\Delta^2(p, T) = m_0^2(p) + \Sigma(\Delta, T). \tag{A.4}$$

The first term in the self-energy Eq. (A.3) renormalises the bare mass term m_0^2, such that $m_0^2 + (N+2)\alpha_\Lambda \int \frac{d^3k}{(2\pi)^3}\frac{1}{2\omega_k} \to m_\Lambda^2$ has logarithmic dependence on the energy scale Λ. After RG, this part results in Eq. (2.3). The second term, or the 'temperature perturbation', only contributes to the logarithmic running via its influence on the infrared cutoff. To make these statements more clear, consider zero temperature such that only the first term contributes. We write the two point function as

$$\begin{aligned}\Gamma^{(2)} &= m^2 + (N+2)\alpha_\Lambda \int_\Lambda^{\Lambda_c} \frac{d^3k}{(2\pi)^3}\frac{1}{2\sqrt{k^2+m^2}} \\ &= m^2 - \frac{(N+2)\alpha_\Lambda}{8\pi^2} m^2 \ln\left(\frac{\Lambda_c}{\Lambda}\right).\end{aligned} \tag{A.5}$$

We use the Callan-Symanzik equation to find the (mass) Beta function

$$\begin{aligned}0 &= \left[\frac{d}{d\ln(\Lambda_c/\Lambda)} + \beta_m(\Lambda)\frac{d}{dm^2}\right]\Gamma^{(2)} \\ \Rightarrow \beta_m(\Lambda) &= \frac{(N+2)\alpha_\Lambda m^2}{8\pi^2} \\ \Rightarrow \frac{dm^2}{d\ln(\Lambda_0/\Lambda)} &= -\frac{(N+2)\alpha_\Lambda m^2}{8\pi^2} \\ &= \left(\frac{-(N+2)}{N+8}\right)\frac{\frac{N+8}{8\pi^2}\alpha_0}{1 + \frac{(N+8)\alpha_0}{8\pi^2}\ln(\Lambda_0/\Lambda)}\end{aligned}$$

$$\frac{d\ln(m^2)}{d\ln(\Lambda_0/\Lambda)} = \left(\frac{-(N+2)}{N+8}\right) \frac{\frac{N+8}{8\pi^2}\alpha_0}{1+\frac{(N+8)\alpha_0}{8\pi^2}\ln(\Lambda_0/\Lambda)}$$

$$m_\Lambda^2 = m_0^2 \left(\frac{\alpha_\Lambda}{\alpha_0}\right)^{\frac{N+2}{N+8}} \tag{A.6}$$

Including non-zero temperatures does not change the form of the running coupling nor mass Eqs. (A.2, A.6), but it does shift the infrared cutoff from $m(p) \to \Lambda = \max\{\Delta_t(p, T), T\}$.

After accounting for how the coupling terms m^2 and α depend on the scale, we find that the gap takes the form

$$\Delta_t^2(p, T, \Lambda) = \gamma^2(p_c - p)\left[\frac{\alpha_\Lambda}{\alpha_0}\right]^{\frac{N+2}{N+8}} + (N+2)\alpha_\Lambda \sum_{\mathbf{k}} \frac{1}{\omega_k} \frac{1}{e^{\frac{\omega_k}{T}} - 1}. \tag{A.7}$$

A.3 Self-energy in the Ordered Phase

The ordered phase is induced by the spontaneous breakdown of the O(3) symmetry when $p > p_c$, as discussed in the introduction. It is a delicate task to calculate the self-energy contributions to the Higgs gap, since within the ordered phase our calculations at each order in α must preserve the Goldstone theorem. The Goldstone theorem is a direct result of the remaining O(2) symmetry. We outline the procedure here. In the Lagrangian, the field $\vec{\varphi} = (\vec{\pi}, \varphi_c + \sigma)$ is shifted such that the minimum of the potential is φ_c, and the field oscillations about this shifted minimum are the two Goldstone modes $\vec{\pi}$ and the gapped Higgs mode σ.

We can write an effective potential, $\mathscr{V}$, from the non-derivative terms of the Lagrangian expanded about the the minimum φ_c

$$\mathscr{V} = -\frac{1}{2}|m^2|(\vec{\pi}, \varphi_c + \sigma)^2 + \frac{1}{4}\alpha\left[(\vec{\pi}, \varphi_c + \sigma)^2\right]^2 \tag{A.8}$$

The following two conditions must simultaneously hold true to ensure that φ_c is indeed the minimum of the potential, and that to any order in α, the perturbations respect the O(2) symmetry and so preserve the Goldstone theorem

$$\left.\frac{d\mathscr{V}}{d\vec{\varphi}}\right|_{\varphi_c} = 0, \quad \text{and} \quad \left.\frac{d^2\mathscr{V}}{d\vec{\pi}^2}\right|_{\varphi_c} = 0. \tag{A.9}$$

Since we have already obtained the universal scale dependence of α_Λ and m_Λ, we do not need to repeat the Callan-Symanzik, RG procedure. We just outline how the thermal perturbations are to be treated. Computing the thermal loops explicitly we obtain the first expression

$$\left.\frac{d\mathscr{V}}{d\vec{\varphi}}\right|_{\varphi_c} = \alpha_\Lambda \varphi_c^2 - |m_\Lambda^2| + (N-1)\alpha_\Lambda \sum_{\mathbf{k}} \frac{1/k}{(e^{\frac{k}{T}}-1)} + 3\alpha_\Lambda \sum_{\mathbf{k}} \frac{1/\omega_{\mathbf{k}}}{(e^{\frac{\omega_{\mathbf{k}}}{T}}-1)} = 0 \tag{A.10}$$

$$\varphi_c^2 = \frac{|m_\Lambda^2|}{\alpha_\Lambda} - (N-1)\sum_{\mathbf{k}} \frac{1/k}{(e^{\frac{k}{T}}-1)} - 3\sum_{\mathbf{k}} \frac{1/\omega_{\mathbf{k}}}{(e^{\frac{\omega_{\mathbf{k}}}{T}}-1)} \tag{A.11}$$

where the thermal corrections are split into two separate contributions. This is because one type comes from the one loop self-energy with a Higgs propagator, and the other with a Goldstone propagator. The first summation accounts for loops with massless Goldstone propagators, while the second accounts for loops with massive Higgs propagators, so that $\omega_{\boldsymbol{k}}^2 = \boldsymbol{k}^2 + \Delta_H(p,T)^2$. We can now find the Higgs gap using the result from Eq. (A.11). Directly computing the one loop corrections to the Higgs gap, we find

$$\begin{aligned}\Delta_H^2 &= 3\alpha_\Lambda \varphi_c^2 - |m_\Lambda^2| + (N-1)\alpha_\Lambda \sum_{\mathbf{k}} \frac{1/k}{(e^{\frac{k}{T}}-1)} + 3\alpha_\Lambda \sum_{\mathbf{k}} \frac{1/\omega_{\mathbf{k}}}{(e^{\frac{\omega_{\mathbf{k}}}{T}}-1)} \\ &= 2|m_\Lambda|^2 - 2(N-1)\alpha_\Lambda \sum_{\mathbf{k}} \frac{1/k}{(e^{\frac{k}{T}}-1)} - 6\alpha_\Lambda \sum_{\mathbf{k}} \frac{1/\omega_{\mathbf{k}}}{(e^{\frac{\omega_{\mathbf{k}}}{T}}-1)},\end{aligned} \tag{A.12}$$

and we have used Eq. (A.11) in passing from the first to second lines. We see that $\Delta_H^2 = 2\alpha_\Lambda\varphi_c^2 + O(\alpha^2)$.

A.4 Néel Temperature

Approaching from the disordered phase, we calculate the Néel temperature by solving Eq. (A.7) for $\Delta_t(p, T_N) = 0$

$$T_N(p)^2 = \frac{\gamma^2(p-p_c)}{(N+2)\alpha_0 \sum_{\mathbf{y}} \frac{1/\omega_y}{(e^{\omega_y}-1)}} \left[\frac{\alpha_0}{\alpha_\Lambda}\right]^{\frac{6}{N+8}}, \tag{A.13}$$

where $\omega_y = \sqrt{\mathbf{y}^2 + (\Gamma/T_N)^2} = \sqrt{\mathbf{y}^2 + \xi^2}$. The fit $\Gamma = \xi T$ was discussed in Chap. 2. Similarly, we can approach from the ordered phase and calculate the Néel temperature by solving Eq. (A.12) for $\Delta_H(p, T_N) = 0$,

$$T_N(p)^2 = \frac{\gamma^2(p-p_c)}{3\alpha_0 \sum_{\mathbf{y}} \frac{1/y}{(e^{\tilde{\omega}_y}-1)} + (N-1)\alpha_0 \sum_{\mathbf{y}} \frac{1/y}{(e^{y}-1)}} \left[\frac{\alpha_0}{\alpha_\Lambda}\right]^{\frac{6}{N+8}}, \tag{A.14}$$

here $\tilde{\omega}_y = \sqrt{\mathbf{y}^2 + \zeta^2}$, and the two terms in the denominator are due to the Higgs and Goldstone self-energies. Since the phase transition is of second order, Eqs. (A.13) and

(A.14) are equivalent. Clearly without account of the finite line width Γ, the equations are identical. As discussed in Chap. 2, we approximate $\Gamma_t = \xi T$ and $\Gamma_H = \zeta T$, from the experimental data. This approximation only becomes important in the vicinity of the phase transition $\Delta_{t/H} < \Gamma_{t/H}$. Equating Eqs. (A.13) and (A.14), we find how ξ and ζ are related at the phase transition. Again, we take $\xi \approx 0.15$ and $\zeta \approx 0.3$.

A.5 Influence of the Spin-Orbit Anisotropy

There is an easy plane anisotropy in $TlCuCl_3$ which can be described by the following additional term in the Lagrangian (2.1)

$$\delta L = -\frac{1}{2} m_x^2 \varphi_x^2 \, . \tag{A.15}$$

The corresponding gap is $m_x = 0.38$ meV [1], implying that the $O(3)$ field theory described in Chap. 2 is valid at the energy scale $\Lambda > m_x$. Far below this scale, $\Lambda \ll m_x$, the theory is effectively $O(2)$.

In Fig. A.1, black lines are identical to that presented in Chap. 2. They are obtained using Eqs. (2.2), (2.5), and (2.7) with $N = 3$ to fit the data. Dashed red lines in Fig. A.1 show the fit of the data using the same equations, but with $N = 2$. Of course the fitting parameters for $N = 2$ are different from that in Eq. (2.8). We find them to be,

$$p_c = 1.02 \text{ kbar}, \ \gamma = 0.675 \text{ meV/kbar}^{1/2}, \ \frac{\alpha_0}{8\pi} = 0.3 \, . \tag{A.16}$$

Solid black and dashed red lines in Fig. A.1 are barely distinguishable. This is because in all equations, N stands next to a large number, see for instance Eq. (2.2). We have also performed a more sophisticated RG calculation, which shows that for $m_x < \Lambda < \Lambda_0$ RG runs with $N = 3$, while for $0 < \Lambda < m_x$, it runs with $N = 2$. We omit the details of this technical calculation and just present the resulting fitting curves in Fig. A.1. The result is shown by the blue dotted lines, and the fitting parameters are

$$p_c = 1.04 \text{ kbar}, \ \gamma = 0.675 \text{ meV/kbar}^{1/2}, \ \frac{\alpha_0}{8\pi} = 0.25 \, . \tag{A.17}$$

Figure A.1 clearly demonstrates that the anisotropy does not influence our conclusions.

It is instructive also to see how accurately the data reproduce the critical index $\nu = (N+2)/(N+8) = 0.455$ in Eqs. (2.5) and (2.7). To do so we consider ν as an independent parameter keeping at the same time $N = 3$ in Eq. (2.2) and in pre-factors in Eqs. (2.5) and (2.7). The fitting curves with $\nu = 0.455$ (solid black), $\nu = 0.36$ (dashed red), $\nu = 0.55$ (dotted blue) are shown in Fig. A.2. From here we conclude that approximately $\nu = 0.45 \pm 0.1$. We do not present a statistical significance of our fits. The point is that often the gap data in experimental papers are given without

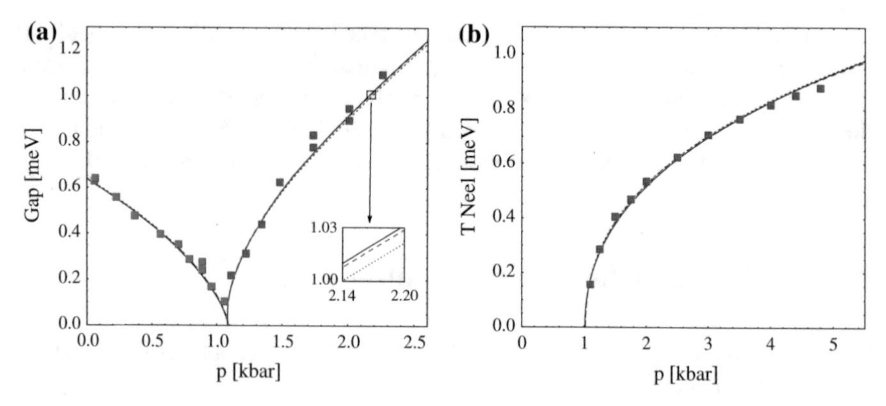

Fig. A.1 Influence of anisotropy on **a** gaps and **b** Néel temperature. Solid black lines, identical to those in Figs. 2.3 and 2.4, are obtained from Eqs. (2.2), (2.5), and (2.7), with $N = 3$. Dashed red lines are obtained from Eqs. (2.2), (2.5), and (2.7), with $N = 2$. Dotted blue lines are obtained by taking into account the anisotropy, which amounts to having a scale dependent N; $N = 3 \rightarrow N = 2$

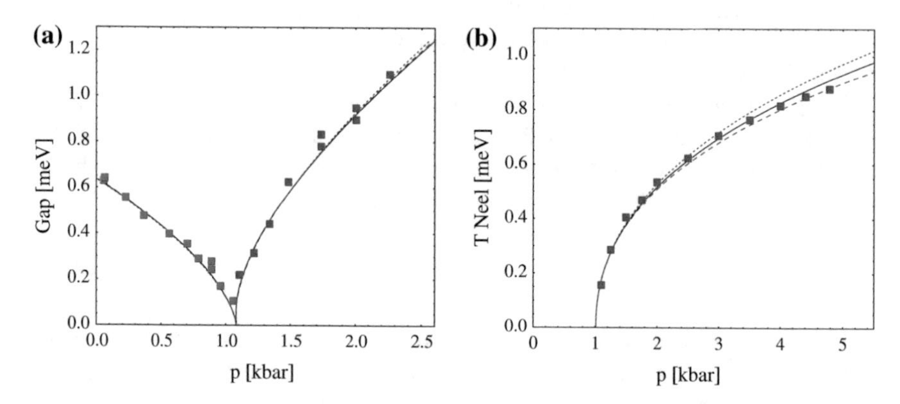

Fig. A.2 Determination of the critical indices for the **a** gaps and **b** Néel temperature. Critical exponent ν replaces the exponent $(N + 2)/(N + 8)$ as seen in Eqs. (2.5) and (2.7). The solid black lines, identical to those in Figs. 2.3 and 2.4, are obtained with $\nu = 0.455$. The dashed red and dotted blue lines are obtained with $\nu = 0.36$ and $\nu = 0.55$, respectively

error bars, even when the error bars are given (say for the Néel temperature) they do not represent statistical error bars. Nevertheless, if we very naively consider the error bars on Néel temperature data as being statistical we conclude that the black curve has $\chi^2 \approx 1$ per degree of freedom, while the red and the blue curves have $\chi^2 \approx 1.5$ and $\chi^2 \approx 3.5$. Where the error bars, not shown in Figs. A.1 or A.2, are $\sigma \approx 0.2$ K ≈ 0.02 meV [2].

Appendix B
Appendix for Chapter 3

This appendix provides details of the bond-operator representation utilised in Chap. 3 to derive various parameters.

B.1 Bond Operator Representation

We employ the bond-operator representation of spin $\boldsymbol{S}$. Define spin at left and right position of bond, $\boldsymbol{S}_l$ and $\boldsymbol{S}_r$. Spins with $S = 1/2$ must satisfy $SU(2)$ algebra,

$$\begin{aligned} [S_{m,\alpha}, S_{m,\beta}] &= i\epsilon_{\alpha,\beta,\gamma} S_{m,\gamma} \,, & [S_{l,\alpha}, S_{r,\beta}] &= 0 \,, \\ \boldsymbol{S}_l \cdot \boldsymbol{S}_r &= -\frac{3}{4} s^\dagger s + \frac{1}{4} t_\alpha^\dagger t_\alpha \,, & S_l^2 = S_r^2 &= \frac{3}{4} \,. \end{aligned} \tag{B.1}$$

Impose constraint $s^\dagger s + t_\alpha^\dagger t_\alpha = 1$ via Lagrange multiplier. Non-derivative/static part of Hamiltonian written in bond operators immediately follows,

$$H_0 = J' \sum_i -\frac{3}{4} s_i^\dagger s_i + \frac{1}{4} t_{i,\alpha}^\dagger t_{i,\alpha} - \mu_i (s_i^\dagger s_i + t_{i,\alpha}^\dagger t_{i,\alpha} - 1) \tag{B.2}$$

subscript i on μ_i makes this a cite dependent chemical potential which accounts for hard-core constraint. Substitution gives the higher-order terms, we keep only the quadratic part for the present discussion

$$H_2 = \frac{J}{2} \sum_{<i,j>} s_i^\dagger s_j^\dagger t_{i,\alpha} t_{j,\alpha} + s_i^\dagger s_j t_{i,\alpha} t_{j,\alpha}^\dagger + \text{H.c.} \tag{B.3}$$

For a mean-field treatment, consider Bose-condensation of singlets and replace, $\langle s^\dagger \rangle = \langle s \rangle = \bar{s}$.

H. Scammell, *Interplay of Quantum and Statistical Fluctuations in Critical Quantum Matter*, Springer Theses,
https://doi.org/10.1007/978-3-319-97532-0

B.2 Fourier and Bogolyubov Transformations

Perform standard Fourier transform, $t_{i,\alpha}^{\dagger} = \frac{1}{\sqrt{N'}} \sum_{\boldsymbol{k}} t_{\boldsymbol{k},\alpha} e^{-i\boldsymbol{k}\cdot\boldsymbol{R}_i}$, with $N' = N/2$ the number of dimers. The quadratic term becomes,

$$\begin{aligned}\bar{H}_2 &= \sum_{\boldsymbol{k}} A_{\boldsymbol{k}} t_{\boldsymbol{k},\alpha}^{\dagger} t_{\boldsymbol{k},\alpha} + \frac{1}{2} B_{\boldsymbol{k}} [t_{\boldsymbol{k},\alpha}^{\dagger} t_{-\boldsymbol{k},\alpha}^{\dagger} + \text{H.c.}] \\ &= \sum_{\boldsymbol{k}} \Omega_{\boldsymbol{k}} \beta_{\boldsymbol{k},\alpha}^{\dagger} \beta_{\boldsymbol{k},\alpha} \,.\end{aligned} \tag{B.4}$$

and the final result is obtained from the Bogolyubov transformation,

$$\begin{aligned} t_{\boldsymbol{k},\alpha}^{\dagger} &= u_{\boldsymbol{k}} \beta_{\boldsymbol{k},\alpha}^{\dagger} - v_{\boldsymbol{k}} \beta_{-\boldsymbol{k},\alpha} \,, & \Omega_{\boldsymbol{k}} &= \sqrt{A_{\boldsymbol{k}}^2 - B_{\boldsymbol{k}}^2} \,, \\ u_{\boldsymbol{k}}^2 / v_{\boldsymbol{k}}^2 &= \pm \frac{1}{2} + \frac{A_{\boldsymbol{k}}}{2\Omega_{\boldsymbol{k}}} \,, & u_{\boldsymbol{k}} v_{\boldsymbol{k}} &= \frac{B_{\boldsymbol{k}}}{2\Omega_{\boldsymbol{k}}} \,. \end{aligned} \tag{B.5}$$

Considering explicitly the geometry of the double cubic lattice model one obtains,

$$\begin{aligned} A_{\boldsymbol{k}} &= \frac{J'}{4} - \mu + J\bar{s}^2[\cos(k_x) + \cos(k_y) + \cos(k_z)] \,, \\ B_{\boldsymbol{k}} &= J\bar{s}^2[\cos(k_x) + \cos(k_y) + \cos(k_z)] \,. \end{aligned} \tag{B.6}$$

B.3 Mean-Field Solution and Parameters; μ, $\bar{s}$

The parameters, μ, $\bar{s}$ are found by the saddle point conditions,

$$\left\langle \frac{\partial H_{MF}}{\partial \mu} \right\rangle = 0 \,, \qquad \left\langle \frac{\partial H_{MF}}{\partial \bar{s}} \right\rangle = 0 \,, \tag{B.7}$$

with $H_{MF} = \bar{H}_0 + \bar{H}_2$. It is convenient to introduce the dimensionless parameter, d,

$$d = \frac{2J\bar{s}^2}{\frac{J'}{4} - \mu} \tag{B.8}$$

which results in the following self-consistent equations,

$$d = \frac{J}{J'} \left(5 - \frac{3}{N'} \sum_{\boldsymbol{k}} \frac{1}{\sqrt{1 + 2d\gamma_{\boldsymbol{k}}}} \right) \,,$$

$$\bar{s}^2 = \frac{5}{2} - \frac{3}{2N'} \sum_{\boldsymbol{k}} \frac{1 + d\gamma_{\boldsymbol{k}}}{\sqrt{1 + 2d\gamma_{\boldsymbol{k}}}} ,$$

$$\mu = -\frac{3J'}{4} + \frac{3J}{N'} \sum_{\boldsymbol{k}} \frac{\gamma_{\boldsymbol{k}}}{\sqrt{1 + 2d\gamma_{\boldsymbol{k}}}} ,$$

$$\gamma_{\boldsymbol{k}} = \frac{1}{2}[\cos(k_x) + \cos(k_y) + \cos(k_z)] . \tag{B.9}$$

The spectrum and gap immediately follow,

$$\Omega_{\boldsymbol{k}} = \left(\frac{J'}{4} - \mu\right)[1 + 2d\gamma_{\boldsymbol{k}}]^{1/2} ,$$

$$\Delta_{(\pi,\pi,\pi)} = \left(\frac{J'}{4} - \mu\right)[1 - 3d]^{1/2} . \tag{B.10}$$

Appendix C
Appendix for Chapter 4

C.1 Real Part of the Self-energy: Non-RG Contribution

In Chap. 4 we self-consistently solve the golden rule of quantum kinetics, Eqs. (4.10), (4.15), to find the imaginary part of the self-energy as well as the structure factor. In doing so we ignore the small frequency dependence of the real part of the self-energy, $\Re\Sigma_q(\omega)$. Our approximation is equivalent to taking $\Re\Sigma_q(\omega) \approx \Re\Sigma_q(\Delta_0)$, where Δ_0 is the physical mass calculated using RG. In this appendix we take into account the full frequency dependence of the real part of the self-energy. This is achieved by adding the frequency dependent correction to the mass gap, $\delta\Sigma(\omega) \equiv \Re\Sigma_q(\omega) - \Re\Sigma_q(\Delta_0)$, and solving the following set of equations self-consistently,

$$\Delta^2(\omega) = \Delta_0^2 + \delta\Sigma(\omega)\,, \tag{C.1}$$

$$\Gamma_q(\omega) = -\frac{\Im\Sigma_q(\omega)}{\omega}\,, \tag{C.2}$$

$$A_q(\omega) = \frac{1}{\pi}\left\{\frac{\omega\Gamma_q(\omega)}{[\omega^2 - \left(q^2 + \Delta^2(\omega)\right)]^2 + \omega^2\Gamma_q^2(\omega)}\right\}. \tag{C.3}$$

Here $\Gamma_q(\omega)$ is defined as in Eq. (4.15), the spectral density $A_q(\omega) \equiv (1 - e^{-\omega/T})\, S_q(\omega)$, while the real part is found via analytic properties (Kramers-Kronig relation),

$$\begin{aligned}\Re\Sigma_q(\omega, T) &= \frac{1}{\pi}\mathscr{P}\int_{-\infty}^{+\infty}\frac{\Im\Sigma_q(\omega', T)}{\omega' - \omega}d\omega' \\ &= \frac{1}{\pi}\mathscr{P}\int_{-\infty}^{+\infty}\frac{-\omega'\Gamma_q(\omega')}{\omega' - \omega}d\omega'\,. \end{aligned}\tag{C.4}$$

Here we ignore momentum dependence, which would give some small additional correction. Since we already know $\Gamma_q(\omega)$ from solving the golden rule of quantum kinetics, we can use the Kramers-Kronig relation Eq. (C.4) to evaluate the real part.

H. Scammell, *Interplay of Quantum and Statistical Fluctuations in Critical Quantum Matter*, Springer Theses,
https://doi.org/10.1007/978-3-319-97532-0

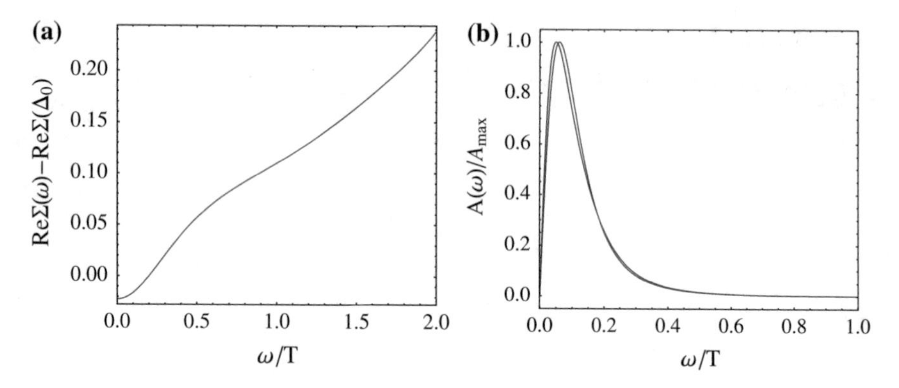

Fig. C.1 **a** Frequency dependent correction to mass gap; the non-RG contribution to the real part of self-energy. **b** The (normalised) spectral density $A_{q=0}(\omega)$: (Blue curve) Including the non-RG, frequency dependent correction. (Maroon curve) excluding the non-RG, frequency dependent correction

The results are shown in Fig. C.1 for the data point $\{\Delta_0, T\} = \{0.2, 0.5\}$ meV, with coupling constant $\beta = 0.15$. Figure C.1a shows the frequency dependence of the non-RG contribution to the real part of the self-energy. Figure C.1b shows the spectral density with and without inclusion of the frequency dependent real part of self-energy, blue and maroon curves, respectively. We see that the inclusion of the real part has a negligible influence.

Appendix D
Appendix for Chapter 5

D.1 Perturbation Theory, Infrared Divergences and Dimensional Reduction: Comments

The technical complication with dimensional reduction is that standard field theoretic RG approaches *fail*—the presence of non-zero temperature introduces infrared divergences into certain classes of Feynman diagrams as the temperature approaches T_N, see Chap. 36 of Ref. [3]. Due to the infrared divergences, it becomes a delicate task to describe observables across the entire range $0 \leq T \leq T_N$ within standard RG techniques, instead more sophisticated approaches must be developed. In this appendix we introduce important aspects of the modified perturbation theory employed in Chap. 5. We also refer the reader to two other developments that relate to the present work, yet with some important distinctions. We call them Approach I and II. Approach I has been developed in a series of papers by O'Connor and Stephens [4–9], and Approach II due to Sachdev [10]. Both methods are modifications of the ϵ-expansion whereby $d = 4 - \epsilon$, and are hence Euclidean field theories (imaginary frequency). Moreover, Approaches I and II have been developed as theories below the upper critical dimension, $d_c = 4$, and as such, logarithmic corrections are not present. In contrast, the approach of the present work (Chap. 5) is explicitly developed for the upper critical dimension. And we note that the logarithmic corrections, due to being at the upper critical dimension, are an important feature of our results in Chap. 5.

D.2 Modified Perturbation Theory

As mentioned, the introduction of non-zero temperature into the $3 + 1$D quantum system renders a certain class of Feynman diagrams are infrared divergent. In fact, such infrared divergences are not specific to $3 + 1$D quantum systems, instead they

H. Scammell, *Interplay of Quantum and Statistical Fluctuations in Critical Quantum Matter*, Springer Theses,
https://doi.org/10.1007/978-3-319-97532-0

are rather generic, see e.g. [3]. In any case, we are solely interested in the problem of non-zero temperatures in a 3 + 1D quantum system. One such class of Feynman diagrams exhibiting infrared divergences are those containing the double internal propagator contributions,

$$T\sum_n \int \frac{d^3k}{(2\pi)^3} \frac{1}{(\omega_n^2 + \boldsymbol{k}^2 + \Delta^2)^2} \approx \frac{1}{8\pi}\frac{T}{\Delta}, \tag{D.1}$$

here we evaluate for zero external momenta running through the loop, and in this approximation we take only the $n = 0$ Matsubara frequency. This is fine for our purposes since for $\Delta \ll T$ this contribution dominates the summation, and summation over all remaining $n \neq 0$ modes is infrared convergent. Diagrams containing such contributions appear at second-order in the perturbative expansion in coupling constant α, and are shown in Fig. D.1. The infrared divergent contributions appear in both the expansion for the coupling constant and mass gap, Fig. D.1a and b, and ultimately are responsible for the difficulty in providing a systematic description of dimensional reduction. To see the issue, let us detail the perturbative corrections to the coupling constant arising from the diagrams shown in Fig. D.1 (ignoring external momenta),

$$\begin{aligned} \alpha &= \alpha_0 - (N+8)\alpha_0^2 T \sum_n \int \frac{d^3k}{(2\pi)^3} \frac{1}{(\omega_n^2 + \boldsymbol{k}^2 + \Delta^2)^2}, \\ &\approx \alpha_0 - \frac{(N+8)\alpha_0^2}{8\pi}\frac{T}{\Delta}. \end{aligned} \tag{D.2}$$

In passing to the second line we again take the $n = 0$ frequency, this is just to most easily demonstrate our point. The coupling constant then appears to be infrared divergent for $\Delta/T \to 0$, i.e. at the Néel temperature.

We will only explicitly discuss the infrared divergence in the running coupling α_Λ and illustrate how it is tamed within our approach, as developed in Chap. 2. The crucial step in our formalism is to insert IR scale $\max[\Delta, T]$ in running coupling. A standard perturbative expansion would obtain temperature dependent IR divergences at second-order in α. One does not expect IR divergences to be physical—they ought to be dealt with in some renormalisation scheme. Moreover, for a scale varying from $\Delta(T=0)$ to $T = T_N$, with $\Delta(0) \sim T_N$, one does not expect α to vary significantly since the relevant energy scales have not varied significantly. This physical expectation is consistent with our scheme.

It is now our task to show, using a Callan-Symanzik type approach, that α runs with scale T, as has been conjectured in previous chapters. We begin with the four point function, taking into account non-zero temperatures via the Matsubara summation,

$$\Gamma^{(4)} = \alpha - (N+8)\alpha^2 T \sum_n \int \frac{d^3k}{(2\pi)^3} \frac{1}{(\omega_n^2 + \boldsymbol{k}^2 + \Delta^2)^2}. \tag{D.3}$$

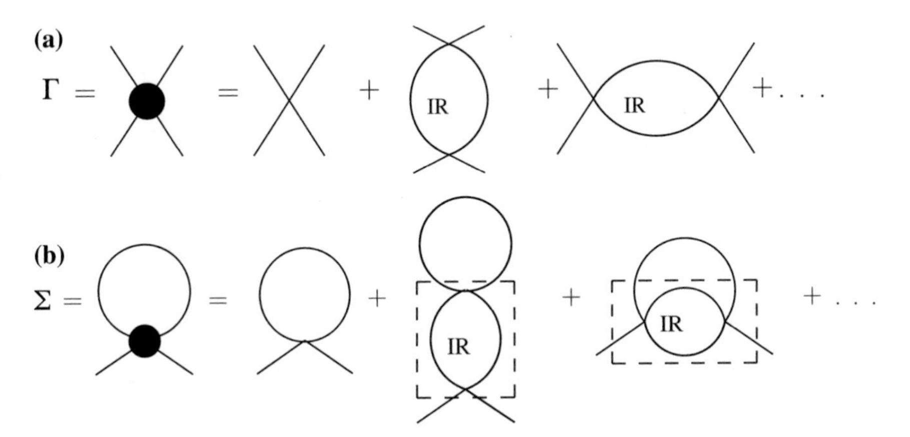

Fig. D.1 Diagrammatic subseries: **a** Vertex, and **b** Self-energy. The infrared divergent loop contributions are marked with "IR"

Applying the Callan-Symanzik equation one obtains,

$$\Lambda\frac{\partial\alpha}{\partial\Lambda} = \frac{(N+8)\alpha^2}{2\pi^2}T\sum_n \frac{\Lambda^3}{((2n\pi T)^2+\Lambda^2)^2}, \tag{D.4}$$

$$\alpha_\Lambda = \frac{\alpha_0}{1+\frac{(N+8)\alpha_0}{2\pi^2}\int_{2\pi\Lambda}^{2\pi\Lambda_0} d\Lambda' T\sum_n \frac{\Lambda'^2}{((2n\pi T)^2+\Lambda'^2)^2}}. \tag{D.5}$$

Finally, setting $\Lambda = T$, one easily verifies (e.g. numerically) that the running coupling in Eq. (D.5) is equivalent to the result derived in Chap. 2, namely,

$$\alpha_T = \frac{\alpha_0}{1+\frac{(N+8)\alpha_0}{8\pi^2}\ln\left(\frac{\Lambda_0}{T}\right)}. \tag{D.6}$$

The point of this analysis being that taking the infrared cut-off to be set by the temperature T, in the vicinity of T_N, acts to cure the unphysical infrared divergences.

Appendix E
Appendix for Chapters 6 and 7

E.1 Greens Functions and Perturbative Decoupling: Disordered Phase

In the disordered phase we have the Lagrangian,

$$\mathscr{L} = \frac{1}{2}\vec{\varphi}^T \hat{G}_D^{-1}\vec{\varphi} - \frac{\alpha}{4}\vec{\varphi}^4 , \tag{E.1}$$

$$\hat{G}_D^{-1} = \begin{pmatrix} \omega^2 - k^2 - (m^2 - B^2) & 2i\omega B & 0 \\ -2i\omega B & \omega^2 - k^2 - (m^2 - B^2) & 0 \\ 0 & 0 & \omega^2 - k^2 - m^2 \end{pmatrix} . \tag{E.2}$$

We choose to work with real fields, in the Cartesian basis $\vec{\varphi} = (\varphi_x, \varphi_y, \varphi_z)$ and therefore do not diagonalise the kinetic matrix, $\hat{G}_D^{-1}$. We are therefore left with anomalous Greens functions, G_{xy}, G_{yx}, but they do not contribute to loop corrections to first-order in α_0. The matrix $\hat{G}_D$ contains the bare Greens functions. We use an effective potential denoted by $\mathscr{V}$, which is the part of the Lagrangian (E.1) independent of derivatives. Then, using a Wick decoupling of the interaction term $\frac{1}{4}\alpha_0\vec{\varphi}^{\,4}$, to first-order in α_0 we find

$$\begin{aligned} \frac{\partial^2 \mathscr{V}}{\partial \varphi_x^2} &= m_0^2 - B^2 + 3\alpha_0\langle\varphi_x^2\rangle + \alpha_0\langle\varphi_y^2\rangle + \alpha_0\langle\varphi_z^2\rangle \\ \frac{\partial^2 \mathscr{V}}{\partial \varphi_y^2} &= m_0^2 - B^2 + \alpha_0\langle\varphi_x^2\rangle + 3\alpha_0\langle\varphi_y^2\rangle + \alpha_0\langle\varphi_z^2\rangle \\ \frac{\partial^2 \mathscr{V}}{\partial \varphi_z^2} &= m_0^2 + \alpha_0\langle\varphi_x^2\rangle + \alpha_0\langle\varphi_y^2\rangle + 3\alpha_0\langle\varphi_z^2\rangle \end{aligned} \tag{E.3}$$

where $\langle\varphi_x^2\rangle$ is the loop integral over the Greens function, G_{xx}, of the φ_x field. The bare Greens functions follow immediately from (E.2), they are

H. Scammell, *Interplay of Quantum and Statistical Fluctuations in Critical Quantum Matter*, Springer Theses,
https://doi.org/10.1007/978-3-319-97532-0

$$
\begin{aligned}
G_{xx}(\omega, \boldsymbol{k}) = G_{yy}(\omega, \boldsymbol{k}) &= \frac{\omega^2 - k^2}{(\omega^2 - (\omega_{\boldsymbol{k}}^{+})^2)(\omega^2 - (\omega_{\boldsymbol{k}}^{-})^2)} \\
G_{zz}(\omega, \boldsymbol{k}) &= \frac{1}{(\omega^2 - (\omega_{\boldsymbol{k}}^{0})^2)} \\
G_{yx}(\omega, \boldsymbol{k}) = G_{xy}^{*}(\omega, \boldsymbol{k}) &= \frac{2i\omega B}{(\omega^2 - (\omega_{\boldsymbol{k}}^{+})^2)(\omega^2 - (\omega_{\boldsymbol{k}}^{-})^2)},
\end{aligned}
\tag{E.4}
$$

with dispersions $\omega_{\boldsymbol{k}}^{\sigma}$ as defined in Eq. (6.2).

E.2 Running Coupling Constant

The four point vertex is calculated to second order in α, the infrared cut-off, Λ, is given by the mass, the magnetic field or the temperature scale; $\max\{m_{\Lambda}, B, T\}$. We use a Callan-Symanzik equation to find the Beta function

$$
\begin{aligned}
\Gamma^{(4)} &= \alpha - 11\alpha^2 \int_{\Lambda}^{\Lambda_c} \frac{d^4k}{(2\pi)^4}\frac{1}{k^4} \\
0 &= \left[\frac{d}{d\ln(\Lambda_c/\Lambda)} + \beta(\alpha)\frac{d}{d\alpha}\right]\Gamma^{(4)} \\
\alpha_{\Lambda} &= \frac{\alpha_0}{1 + \frac{11\alpha_0}{8\pi^2}\ln(\Lambda_0/\Lambda)}
\end{aligned}
\tag{E.5}
$$

where Λ_c is some momentum cut-off such as the inverse lattice spacing, while Λ_0 is the normalisation point.

E.3 Running Mass

Consider the corrections to the curvature (E.3), under renormalisation we replace the bare coupling with the running running coupling $\alpha_0 \to \alpha_{\Lambda}$, and now explicitly substituting loop integrals (with $i = \{x, y\}$)

$$
\begin{aligned}
\frac{\partial^2 \mathscr{V}}{\partial \varphi_i^2} &= m_0^2 - B^2 + 5\alpha_{\Lambda} \int \frac{d^3k}{(2\pi)^3}\frac{1}{2\omega_{\boldsymbol{k}}^0} + \alpha_{\Lambda} \int \frac{d^3k}{(2\pi)^3}\frac{1}{\omega_{\boldsymbol{k}}^0}\{2n(\omega_{\boldsymbol{k}}^{+}) + 2n(\omega_{\boldsymbol{k}}^{-}) + n(\omega_{\boldsymbol{k}}^{0})\}\,, \\
\frac{\partial^2 \mathscr{V}}{\partial \varphi_z^2} &= m_0^2 + 5\alpha_{\Lambda} \int \frac{d^3k}{(2\pi)^3}\frac{1}{2\omega_{\boldsymbol{k}}^0} + \alpha_{\Lambda} \int \frac{d^3k}{(2\pi)^3}\frac{1}{\omega_{\boldsymbol{k}}^0}\{n(\omega_{\boldsymbol{k}}^{+}) + n(\omega_{\boldsymbol{k}}^{-}) + 3n(\omega_{\boldsymbol{k}}^{0})\}\,.
\end{aligned}
\tag{E.6}
$$

The coupling constant coefficient is the running coupling α_{Λ} since the two point corrections are multiplicative with the four point vertices. The integral first terms in (E.6) renormalise the bare mass term m_0^2, such that $m_0^2 + 5\alpha_{\Lambda} \int \frac{d^3k}{(2\pi)^3}\frac{1}{2\omega_k} \to m_{\Lambda}^2$

has logarithmic dependence on the energy scale Λ. The second integral terms, or the 'thermal perturbations', only contributes to the logarithmic running via its influence on the infrared cutoff. To make these statements more clear, consider zero temperature such that only the first term contributes. We write the two point function as (with forward substitution of running mass $m_0 \to m_\Lambda$

$$\begin{aligned}\Gamma^{(2)} \equiv \frac{\partial^2 \mathscr{V}}{\partial \varphi_z^2}\Big|_{T=0} &= m_\Lambda^2 + 5\alpha_\Lambda \int_\Lambda^{\Lambda_c} \frac{d^3k}{(2\pi)^3} \frac{1}{2\sqrt{k^2 + m_\Lambda^2}} \\ &= m_\Lambda^2 - \frac{5\alpha_\Lambda}{8\pi^2} m_\Lambda^2 \ln\left(\frac{\Lambda_c}{\Lambda}\right) \\ &= \frac{\partial^2 \mathscr{V}}{\partial \varphi_i^2}\Big|_{T=0} + B^2. \end{aligned} \tag{E.7}$$

We note that the logarithmic correction is independent of magnetic field B, which is essential to ensure that B is not renormalised. We use the Callan-Symanzik equation to find the (mass) Beta function

$$\begin{aligned} 0 &= \left[\frac{d}{d\ln(\Lambda_c/\Lambda)} + \beta_m(\Lambda)\frac{d}{dm_\Lambda^2}\right]\Gamma^{(2)} \\ \beta_m(\Lambda) &= \frac{5\alpha_\Lambda m_\Lambda^2}{8\pi^2} \\ \frac{dm_\Lambda^2}{d\ln(\Lambda_0/\Lambda)} &= -\frac{5\alpha_\Lambda m_\Lambda^2}{8\pi^2} \\ m_{\Lambda,\sigma}^2 &= m_0^2\left(\frac{\alpha_\Lambda}{\alpha_0}\right)^{\frac{5}{11}} \end{aligned} \tag{E.8}$$

In this last line we explicitly give an index σ to denote the different polarisations. At zero temperature, the terms $m_{\Lambda,\sigma}$ are equivalent for all polarisations, σ. Including non-zero temperatures does not change the form of the running coupling nor mass Eqs. (E.5), (E.8), but it does: (i) influence the infrared cutoff from $\Lambda = \max\{m_\Lambda, B, T\}$; and (ii) lift the degeneracy of the mass terms, which now explicitly become,

$$m_{\Lambda,\pm}^2 = m_0^2\left[\frac{\alpha_\Lambda}{\alpha_0}\right]^{\frac{5}{11}} + \alpha_\Lambda \sum_k \frac{1}{\omega_k^0}\{2n(\omega_{\boldsymbol{k}}^+) + 2n(\omega_{\boldsymbol{k}}^-) + n(\omega_{\boldsymbol{k}}^0)\}\,, \tag{E.9}$$

$$m_{\Lambda,z}^2 = m_0^2\left[\frac{\alpha_\Lambda}{\alpha_0}\right]^{\frac{5}{11}} + \alpha_\Lambda \sum_k \frac{1}{\omega_k^0}\{n(\omega_{\boldsymbol{k}}^+) + n(\omega_{\boldsymbol{k}}^-) + 3n(\omega_{\boldsymbol{k}}^0)\}\,. \tag{D.5}$$

Here $n(\omega_{\boldsymbol{k}}) = 1/(e^{\frac{\omega_{\boldsymbol{k}}}{T}} - 1)$. It is straightforward to check that expansions of Eqs. (E.9) and (D.5) in powers of B contain only even powers. Interestingly these expansions are different for $m_{\Lambda,\pm}$ and $m_{\Lambda,0}$. Therefore the relation $\omega_{\boldsymbol{k}}^{+} - \omega_{\boldsymbol{k}}^{0} = \omega_{\boldsymbol{k}}^{0} - \omega_{\boldsymbol{k}}^{-}$, which is exact at $T = 0$, does not hold at non-zero T. At non-zero T the relation is valid only up to the linear-in-B approximation.

E.4 Greens Functions in the Ordered Phase

In the ordered phase the Lagrangian takes the form,

$$\mathscr{L} = \frac{1}{2}\vec{\varphi}^T \hat{G}_O^{-1} \vec{\varphi} - \frac{\alpha}{4}\{(\sigma^2 + \pi^2 + z^2)^2 + 4\varphi_c^3\sigma + 4\varphi_c\sigma(\sigma^2 + \pi^2 + z^2)\}\,, \tag{E.10}$$

$$\hat{G}_O^{-1} = \begin{pmatrix} \omega^2 - k^2 - 2(B^2 - m^2) & 2i\omega B & 0 \\ -2i\omega B & \omega^2 - k^2 & 0 \\ 0 & 0 & \omega^2 - k^2 - B^2 \end{pmatrix}, \tag{E.11}$$

where $\vec{\varphi}^T = (\varphi_c + \sigma, \pi, z)$. Similarly to the treatment of the disordered phase, we choose to work with real fields which do not diagonalise the kinetic matrix, $\hat{G}_O^{-1}$. We are therefore left with anomalous Greens functions, e.g. $G_{\sigma\pi}$. The matrix $\hat{G}_O$ contains the bare Greens functions as its entries, they are

$$\begin{aligned}
G_{\pi\pi}(\omega, \boldsymbol{k}) &= \frac{\omega^2 - k^2 - 2(B^2 - m_{\Lambda,H}^2)}{(\omega^2 - (\omega_{\boldsymbol{k}}^H)^2)(\omega^2 - (\omega_{\boldsymbol{k}}^G)^2)} \\
G_{\sigma\sigma}(\omega, \boldsymbol{k}) &= \frac{\omega^2 - k^2}{(\omega^2 - (\omega_{\boldsymbol{k}}^H)^2)(\omega^2 - (\omega_{\boldsymbol{k}}^G)^2)} \\
G_{zz}(\omega, \boldsymbol{k}) &= \frac{1}{(\omega^2 - (\omega_{\boldsymbol{k}}^z)^2)} \\
G_{\pi\sigma}(\omega, \boldsymbol{k}) = G_{\sigma\pi}^*(\omega, \boldsymbol{k}) &= \frac{2i\omega B}{(\omega^2 - (\omega_{\boldsymbol{k}}^H)^2)(\omega^2 - (\omega_{\boldsymbol{k}}^G)^2)}\,,
\end{aligned} \tag{E.12}$$

with dispersions as defined in Eqs. (6.7), (6.8), and (6.9).

E.5 Matsubara Loop Integrals

Finite temperature loop integrals are conveniently carried out utilising the imaginary time Fourier transform [11],

$$\frac{1}{x^2 + E^2} = \int_0^\beta d\tau e^{ix\tau} \frac{1}{2E}\left[(1 + n(E))e^{-E\tau} + n(E)e^{E\tau}\right]\,, \tag{E.13}$$

where $n(E)$ is the usual Bose-occupation factor, $\beta = 1/T$, and a useful identity is $(1/\beta)\sum_n e^{ix(\tau_1-\tau_2)} = \delta(\tau_1 - \tau_2)$. The single loop integrals are

$$\langle\sigma^2\rangle = \frac{1}{\beta}\sum_{n=-\infty}^{\infty}\int\frac{d^3k}{(2\pi)^3}G_{\sigma\sigma}(\omega_n, \boldsymbol{k})$$

$$= \int\frac{d^3k}{(2\pi)^3}\left\{\frac{\left[6B^2 - 2m^2_{\Lambda,H} + \omega^2_H - \omega^2_G\right]}{\omega^2_H - \omega^2_G}\frac{n_H}{2\omega_H} - \frac{\left[6B^2 - 2m^2_{\Lambda,H} - \omega^2_H + \omega^2_G\right]}{\omega^2_H - \omega^2_G}\frac{n_G}{2\omega_G}\right\}$$

$$\langle\pi^2\rangle = \frac{1}{\beta}\sum_{n=-\infty}^{\infty}\int\frac{d^3k}{(2\pi)^3}G_{\pi\pi}(\omega_n, \boldsymbol{k})$$

$$= \int\frac{d^3k}{(2\pi)^3}\left\{\frac{\left[2B^2 + 2m^2_{\Lambda,H} + \omega^2_H - \omega^2_G\right]}{\omega^2_H - \omega^2_G}\frac{n_H}{2\omega_H} - \frac{\left[2B^2 + 2m^2_{\Lambda,H} - \omega^2_H + \omega^2_G\right]}{\omega^2_H - \omega^2_G}\frac{n_G}{2\omega_G}\right\}$$

$$\langle z^2\rangle = \frac{1}{\beta}\sum_{n=-\infty}^{\infty}\int\frac{d^3k}{(2\pi)^3}G_{zz}(\omega_n, \boldsymbol{k}) = \int\frac{d^3k}{(2\pi)^3}\frac{n_z}{\omega_z}. \tag{E.14}$$

Here $\omega_n = 2n\pi/\beta$ are usual Matsubara frequencies, and notation for Bose-occupation factors and dispersions has been simplified in an obvious way. It is also worth noting that the trace over anomalous Greens functions gives zero,

$$\frac{1}{\beta}\sum_{n=-\infty}^{\infty}\int\frac{d^3k}{(2\pi)^3}G_{\sigma\pi}(\omega_n, \boldsymbol{k}) = \frac{1}{\beta}\sum_{n=-\infty}^{\infty}\int\frac{d^3k}{(2\pi)^3}G_{\pi\sigma}(\omega_n, \boldsymbol{k}) = 0\,. \tag{E.15}$$

The double loop integrals have external momenta $(E, \boldsymbol{p})$ running through the loop, we only work with $\boldsymbol{p} = 0$ and eventually set $E = 0$. The double-loop integrals are,

$$\langle\sigma^2 z^2\rangle = \mathrm{Re}\frac{1}{\beta}\sum_{n=-\infty}^{\infty}\int\frac{d^3k}{(2\pi)^3}G_{\sigma\sigma}(\omega_n, \boldsymbol{k})G_{zz}(E - \omega_n, \boldsymbol{k})$$

$$= \mathrm{Re}\int\frac{d^3k}{(2\pi)^3}\frac{(\omega^2_H - \boldsymbol{k}^2)}{\omega^2_H - \omega^2_G}\frac{1}{2\omega_H 2\omega_z}$$

$$\left\{\left[(1+n_H)(1+n_z) - n_H n_z\right]\left[\frac{1}{iE - \omega_H - \omega_z} - \frac{1}{iE + \omega_H + \omega_z}\right]\right.$$

$$\left.+\left[(1+n_H)n_z - n_H(1+n_z)\right]\left[\frac{1}{iE - \omega_H + \omega_z} - \frac{1}{iE + \omega_H - \omega_z}\right]\right\}$$

$$- \mathrm{Re}\int\frac{d^3k}{(2\pi)^3}\frac{(\omega^2_G - \boldsymbol{k}^2)}{\omega^2_H - \omega^2_G}\frac{1}{2\omega_G 2\omega_z}$$

$$\left\{\left[(1+n_G)(1+n_z) - n_G n_z\right]\left[\frac{1}{iE - \omega_G - \omega_z} - \frac{1}{iE + \omega_G + \omega_z}\right]\right.$$

$$\left.+\left[(1+n_G)n_z - n_G(1+n_z)\right]\left[\frac{1}{iE - \omega_G + \omega_z} - \frac{1}{iE + \omega_G - \omega_z}\right]\right\}. \tag{E.16}$$

The other important double-loop diagram comes from the trace $\langle\sigma^2\pi^2\rangle$, this diagram must be included in order to satisfy the Goldstone theorem. Observe now that the anomalous Greens functions play a crucial role; loops composed of two anomalous Greens functions are non-zero. Let us explicitly evaluate,

$$\begin{aligned}
\langle\sigma^2\pi^2\rangle &= \mathrm{Re}\frac{1}{\beta}\sum_{n=-\infty}^{\infty}\int\frac{d^3k}{(2\pi)^3}\left\{G_{\sigma\sigma}(\omega_n,\boldsymbol{k})G_{\pi\pi}(E-\omega_n,\boldsymbol{k})+G_{\sigma\pi}(\omega_n,\boldsymbol{k})G_{\pi\sigma}(E-\omega_n,\boldsymbol{k})\right\}\\
&= \mathrm{Re}\int\frac{d^3k}{(2\pi)^3}\frac{1}{2\omega_H 2\omega_G}\\
&\left\{[(1+n_H)(1+n_G)-n_H n_G]\left[\frac{1}{iE-\omega_H-\omega_G}-\frac{1}{iE+\omega_H+\omega_G}\right]\right.\\
&\left.+[(1+n_H)n_G-n_H(1+n_G)]\left[\frac{1}{iE-\omega_H+\omega_G}-\frac{1}{iE+\omega_H-\omega_G}\right]\right\}.
\end{aligned} \tag{E.17}$$

It is also interesting to note that at zero temperature the contribution due to the trace over $G_{\sigma\pi}(\omega_n,\boldsymbol{k})G_{\pi\sigma}(E-\omega_n,\boldsymbol{k})$ vanishes.

E.6 Continuity of the Modes

An equally important check of the renormalisation procedure employed in this work is that all modes continuously evolve into their counterpart at the phase transition $B=B_c^{\pm}(T)$,

$$\text{(I)}:\ \omega_{\boldsymbol{k}}^{+}=\omega_{\boldsymbol{k}}^{H},\quad \text{(II)}:\ \omega_{\boldsymbol{k}}^{0}=\omega_{\boldsymbol{k}}^{z},\quad \text{(III)}:\ \omega_{\boldsymbol{k}}^{-}=\omega_{\boldsymbol{k}}^{G}. \tag{E.18}$$

Using the critical fields B_c^-, B_c^+ from Eqs. (6.6) and (6.20), in the dispersion relations for the upper Zeeman triplon ($\sigma=+1$) and Higgs modes Eqs. (6.2) and (6.7), respectively, we easily verify the continuity Eq. (E.18)I,

$$(\omega_{\boldsymbol{0}}^{+})^2=\left(\sqrt{m_0^2+\alpha_0(4\langle\varphi_x^2\rangle+\langle\varphi_z^2\rangle)}+B_c\right)^2=4B_c^2 \tag{E.19}$$

$$(\omega_{\boldsymbol{0}}^{H})^2=2(3B_c^2-m_{\Lambda,H}^2)=2\left(3B_c^2-m_0^2-\alpha_0(3\langle\sigma^2\rangle+\langle\pi^2\rangle+\langle z^2\rangle)\right)=4B_c^2. \tag{E.20}$$

Similarly, for the z-field/precession mode, we find

$$(\omega_{\boldsymbol{0}}^{0})^2=m_0^2+\alpha_0\langle\varphi_x^2\rangle+\alpha_0\langle\varphi_y^2\rangle+3\alpha_0\langle\varphi_z^2\rangle=B_c^2+2\alpha_0\langle\varphi_z^2\rangle-2\alpha_0\langle\varphi_x^2\rangle \tag{E.21}$$

$$(\omega_{\boldsymbol{0}}^{z})^2=B_c^2+2\alpha_0\langle z^2\rangle-2\alpha_0\langle\sigma^2\rangle+4\alpha_0^2\varphi_c^2\langle\sigma^2 z^2\rangle \tag{E.22}$$

and hence to verify the continuity Eq. (E.18)II, it suffices to show that

$$2\alpha_0\langle\varphi_z^2\rangle-2\alpha_0\langle\varphi_x^2\rangle=2\alpha_0\langle z^2\rangle-2\alpha_0\langle\sigma^2\rangle+4\alpha_0^2\varphi_c^2\langle\sigma^2 z^2\rangle. \tag{E.23}$$

Evaluating the loop integrals at B_c, the logarithmic corrections (from the vacuum/quantum sector) identically vanish, and it remains to consider the thermal contributions. Substituting the loop integrals and performing straightforward algebra, we confirm (E.23) and hence the condition (E.18)II.

E.7 Mass Parameter Renormlisation

At zero temperature, the z-field is also unrenormalised by logarithmic corrections; the gap is fixed at the Larmor precession frequency, $\partial^2 \mathscr{V}/\partial z^2 = m_{\Lambda,z}^2 = B^2$. All log corrections in Eq. (6.14) exactly cancel. However at non-zero temperature, the precession mode receives renormalisation due to the heat bath, i.e. the non-zero temperature loop integrals in the following expression do not cancel,

$$m_{\Lambda,z}^2 = B^2 + 2\alpha\langle z^2\rangle - 2\alpha\langle\sigma^2\rangle + 4\alpha^2\varphi_c^2\langle\sigma^2 z^2\rangle \to \begin{cases} = B^2, & \text{for } T = 0, \\ \neq B^2, & \text{for } T \neq 0 . \end{cases} \tag{E.24}$$

This is at first a somewhat unexpected result, yet does not provide any inconsistencies within the current treatment.

Now, the other mass parameter appearing in the ordered phase, $m_{\Lambda,H}$ as defined in Eq. (6.12), does receive logarithmic corrections. Upon RG, as described above, the explicit form of the renormalised mass parameter is,

$$\begin{aligned} m_{\Lambda,H}^2 = m_0^2 \left[\frac{\alpha_\Lambda}{\alpha_0}\right]^{\frac{5}{11}} + \alpha_\Lambda \sum_{\mathbf{k}} \Bigg\{ & \frac{2(5B^2 - m_\Lambda^2)}{(\omega_k^H)^2 - (\omega_k^G)^2} \left[\frac{n(\omega_k^H)}{\omega_k^H} - \frac{n(\omega_k^G)}{\omega_k^G}\right] \\ & + 2\left[\frac{n(\omega_k^H)}{\omega_k^H} + \frac{n(\omega_k^G)}{\omega_k^G}\right] + \frac{n(\omega_k^z)}{\omega_k}\Bigg\} + \mathscr{O}(\varphi_c^2\alpha^2) . \end{aligned} \tag{E.25}$$

Appendix F
Appendix for Chapter 8

F.1 Amplitude Factors

In this section we show how to obtain amplitude factors $\mathscr{A}_{\alpha,\boldsymbol{k}}$, $\mathscr{B}_{\alpha,\boldsymbol{k}}$, $\mathscr{D}_{\alpha,\boldsymbol{k}}$, where $\alpha = H, G$, and for convenience we have defined $\mathscr{B}_{\alpha,\boldsymbol{k}} = \mathscr{D}_{\alpha,\boldsymbol{k}}\mathscr{A}_{\alpha,\boldsymbol{k}}$.

Let $a_{\boldsymbol{k}}/a_{\boldsymbol{k}}^\dagger$, $b_{\boldsymbol{k}}/b_{\boldsymbol{k}}^\dagger$, and $c_{\boldsymbol{k}}/c_{\boldsymbol{k}}^\dagger$ be the annihilation/creation operators of the Higgs, Goldstone, and z-modes. Accordingly, the Hamiltonian reads

$$H = \sum_{\boldsymbol{k}} \left[\omega_{\boldsymbol{k}}^H a_{\boldsymbol{k}}^\dagger a_{\boldsymbol{k}} + \omega_{\boldsymbol{k}}^G b_{\boldsymbol{k}}^\dagger b_{\boldsymbol{k}} + \omega_{\boldsymbol{k}}^z c_{\boldsymbol{k}}^\dagger c_{\boldsymbol{k}} \right] + \text{const.} \tag{F.1}$$

Field operators are expressed in terms of the creation and annihilation operators in the following generic form,

$$\begin{aligned}
\sigma(\boldsymbol{x}, t) = \sum_{\boldsymbol{k}} \Big\{ &\mathscr{A}_{H,\boldsymbol{k}}[a_{\boldsymbol{k}} e^{i\omega_{\boldsymbol{k}}^H t - i\boldsymbol{k}\cdot\boldsymbol{x}} + a_{\boldsymbol{k}}^\dagger e^{-i\omega_{\boldsymbol{k}}^H t + i\boldsymbol{k}\cdot\boldsymbol{x}}] \\
&+ \mathscr{A}_{G,\boldsymbol{k}}[b_{\boldsymbol{k}} e^{i\omega_{\boldsymbol{k}}^G t - i\boldsymbol{k}\cdot\boldsymbol{x}} + b_{\boldsymbol{k}}^\dagger e^{-i\omega_{\boldsymbol{k}}^G t + i\boldsymbol{k}\cdot\boldsymbol{x}}] \Big\} \\
\pi_y(\boldsymbol{x}, t) = \sum_{\boldsymbol{k}} \Big\{ &\mathscr{B}_{H,\boldsymbol{k}}[a_{\boldsymbol{k}} e^{i\omega_{\boldsymbol{k}}^H t - i\boldsymbol{k}\cdot\boldsymbol{x}} - a_{\boldsymbol{k}}^\dagger e^{-i\omega_{\boldsymbol{k}}^\pi t + i\boldsymbol{k}\cdot\boldsymbol{x}}] \\
&+ \mathscr{B}_{G,\boldsymbol{k}}[b_{\boldsymbol{k}} e^{i\omega_{\boldsymbol{k}}^G t - i\boldsymbol{k}\cdot\boldsymbol{x}} - b_{\boldsymbol{k}}^\dagger e^{-i\omega_{\boldsymbol{k}}^G t + i\boldsymbol{k}\cdot\boldsymbol{x}}] \Big\} \\
\pi_z(\boldsymbol{x}, t) = \sum_{\boldsymbol{k}} &\frac{1}{\sqrt{2\omega_{\boldsymbol{k}}^z}} [c_{\boldsymbol{k}} e^{i\omega_{\boldsymbol{k}}^z t - i\boldsymbol{k}\cdot\boldsymbol{x}} + c_{\boldsymbol{k}}^\dagger e^{-i\omega_{\boldsymbol{k}}^z t + i\boldsymbol{k}\cdot\boldsymbol{x}}].
\end{aligned} \tag{F.2}$$

From the Lagrangian $\mathscr{L}_2$ (8.2), we obtain the equations of motion

$$\begin{aligned}
0 =& \partial_\mu^2 \sigma - 2B\dot{\pi}_y + 2(B^2 - m^2)\sigma \\
0 =& \partial_\mu^2 \pi_y + 2B\dot{\sigma} \ .
\end{aligned} \tag{F.3}$$

H. Scammell, *Interplay of Quantum and Statistical Fluctuations in Critical Quantum Matter*, Springer Theses,
https://doi.org/10.1007/978-3-319-97532-0

Substituting expressions (F.2) into (F.3) one obtains

$$
\begin{aligned}
-\left((\omega_{\boldsymbol{k}}^{H})^{2}-\boldsymbol{k}^{2}\right)\mathscr{B}_{H} &= 2i\omega_{\boldsymbol{k}}^{H}B\mathscr{A}_{H,\boldsymbol{k}} \\
-\left((\omega_{\boldsymbol{k}}^{G})^{2}-\boldsymbol{k}^{2}\right)\mathscr{B}_{G} &= 2i\omega_{\boldsymbol{k}}^{G}B\mathscr{A}_{G,\boldsymbol{k}}\,.
\end{aligned}
\tag{F.4}
$$

We note that the necessity of expressing the field operators σ, π_y in terms of mixed operators $a_{\boldsymbol{k}}/a_{\boldsymbol{k}}^{\dagger}$, $b_{\boldsymbol{k}}/b_{\boldsymbol{k}}^{\dagger}$ becomes apparent when solving the equations of motion (F.3), namely, without mixing one cannot satisfy these equations. Finally, one finds

$$
\mathscr{B}_{\alpha}=-\frac{2i\omega_{\boldsymbol{k}}^{\alpha}B}{(\omega_{\boldsymbol{k}}^{\alpha})^{2}-\boldsymbol{k}^{2}}\mathscr{A}_{\alpha,\boldsymbol{k}}\equiv\mathscr{D}_{\alpha,\boldsymbol{k}}\mathscr{A}_{\alpha,\boldsymbol{k}}\,. \tag{F.5}
$$

It remains to find $\mathscr{A}_{\alpha,\boldsymbol{k}}$. To do so, one must Legendre transform $\mathscr{L}_2$ to give the following Hamiltonian, note we treat only the hybridised terms σ, π_y, the Hamiltonian for π_z can be treated separately and is trivial,

$$
\begin{aligned}
H[\sigma,\pi_y] &= \frac{1}{2}\int d^3x dt\left[(\partial_\mu\sigma(\boldsymbol{x},t))^2+2(B^2-m^2)\sigma(\boldsymbol{x},t)^2+(\partial_\mu\pi_y(\boldsymbol{x},t))^2\right] \\
&= \frac{1}{2}\sum_{\boldsymbol{k}}\mathscr{A}_{H,\boldsymbol{k}}^{2}\left[(\omega_{\boldsymbol{k}}^{H})^2+\boldsymbol{k}^2+2(B^2-m^2)-((\omega_{\boldsymbol{k}}^{H})^2+\boldsymbol{k}^2)\mathscr{D}_{H,\boldsymbol{k}}^{2}\right]\left[a_{\boldsymbol{k}}^{\dagger}a_{\boldsymbol{k}}+a_{\boldsymbol{k}}a_{\boldsymbol{k}}^{\dagger}\right] \\
&+ \frac{1}{2}\sum_{\boldsymbol{k}}\mathscr{A}_{G,\boldsymbol{k}}^{2}\left[(\omega_{\boldsymbol{k}}^{G})^2+\boldsymbol{k}^2+2(B^2-m^2)-((\omega_{\boldsymbol{k}}^{G})^2+\boldsymbol{k}^2)\mathscr{D}_{G,\boldsymbol{k}}^{2}\right]\left[b_{\boldsymbol{k}}^{\dagger}b_{\boldsymbol{k}}+b_{\boldsymbol{k}}b_{\boldsymbol{k}}^{\dagger}\right] \\
&= \sum_{\boldsymbol{k}}\left[\omega_{\boldsymbol{k}}^{H}a_{\boldsymbol{k}}^{\dagger}a_{\boldsymbol{k}}+\omega_{\boldsymbol{k}}^{G}b_{\boldsymbol{k}}^{\dagger}b_{\boldsymbol{k}}\right]+\text{const.}
\end{aligned}
\tag{F.6}
$$

where the last line is taken from (F.1). Upon equating the appropriate coefficient of $a_{\boldsymbol{k}}^{\dagger}a_{\boldsymbol{k}}$ and $b_{\boldsymbol{k}}^{\dagger}b_{\boldsymbol{k}}$, and after straightforward algebra, one obtains

$$
\mathscr{A}_{\alpha,\boldsymbol{k}}=\sqrt{\frac{\omega_{\boldsymbol{k}}^{\alpha}B^2}{(B^2+m^2)(\omega_{\boldsymbol{k}}^{\alpha})^2+(3B^2-m^2)(2B^2-2m^2+\boldsymbol{k}^2)}}\,.
$$

Appendix G
Appendix for Chapter 9

G.1 Excitation Modes of the Bose-Condensate

It is convenient to define the following time independent spinors

$$u = \begin{pmatrix} \cos\frac{\theta_0}{2} \\ \sin\frac{\theta_0}{2} e^{i\varphi_0} \end{pmatrix} , \quad \bar{u} = \begin{pmatrix} -\sin\frac{\theta_0}{2} \\ \cos\frac{\theta_0}{2} e^{i\varphi_0} \end{pmatrix} . \tag{G.1}$$

Hence the Bose-condensate field (9.6) is $z_0 = A_0 u$. Substitution of ansatz,

$$\delta z = z_+ e^{i\omega t - i\boldsymbol{k}\cdot\boldsymbol{r}} + z_- e^{-i\omega t + i\boldsymbol{k}\cdot\boldsymbol{r}} , \tag{G.2}$$

in Euler-Lagrange equation (9.7) results in the following algebraic equations for the "amplitude" spinors z_+ and z_-

$$\begin{aligned} (-\omega^2 + k^2) z_+ + 2\omega(\vec{\sigma}\cdot\vec{\mathscr{B}}) z_+ + (\mathscr{B}^2 - m^2)(z_-^\dagger u + u^\dagger z_+) u = 0 , \\ (-\omega^2 + k^2) z_- - 2\omega(\vec{\sigma}\cdot\vec{\mathscr{B}}) z_- + (\mathscr{B}^2 - m^2)(z_+^\dagger u + u^\dagger z_-) u = 0 . \end{aligned} \tag{G.3}$$

Projecting each Eq. (G.3), onto both u and $\bar{u}$, and defining,

$$u^\dagger z_+ = a_1 , \qquad a_1 + a_2^* = a_+ \tag{G.4}$$

$$u^\dagger z_- = a_2 , \qquad a_1 - a_2^* = a_- \tag{G.5}$$

$$\bar{u}^\dagger z_+ = b_1 , \qquad b_1 + b_2^* = b_+ \tag{G.6}$$

$$\bar{u}^\dagger z_- = b_2 , \qquad b_1 - b_2^* = b_- \tag{G.7}$$

we get the following matrix equation for the amplitudes a_+, a_-, b_+, b_-,

H. Scammell, *Interplay of Quantum and Statistical Fluctuations in Critical Quantum Matter*, Springer Theses,
https://doi.org/10.1007/978-3-319-97532-0

$$\begin{pmatrix} k^2-\omega^2+2(\mathscr{B}^2-m^2) & 2\omega\mathscr{B}\cos\theta_0 & 0 & -2\omega\mathscr{B}\sin\theta_0 \\ 2\omega\mathscr{B}\cos\theta_0 & k^2-\omega^2 & -2\omega\mathscr{B}\sin\theta_0 & 0 \\ 0 & -2\omega\mathscr{B}\sin\theta_0 & k^2-\omega^2 & -2\omega\mathscr{B}\cos\theta_0 \\ -2\omega\mathscr{B}\sin\theta_0 & 0 & -2\omega\mathscr{B}\cos\theta_0 & k^2-\omega^2 \end{pmatrix} \begin{pmatrix} a_+ \\ a_- \\ b_+ \\ b_- \end{pmatrix} = 0. \tag{G.8}$$

Frequencies of normal modes (9.8) immediately follow from this matrix equation. Amplitude vectors (a_+, a_-, b_+, b_-), corresponding to the normal modes are of the following form

$$|1\rangle \propto \begin{pmatrix} 0 \\ \sin\theta_0 \\ 1 \\ \cos\theta_0 \end{pmatrix}; \quad |2\rangle \propto \begin{pmatrix} b_2 \\ -\cos\theta_0 \\ 0 \\ \sin\theta_0 \end{pmatrix}; \quad |3\rangle \propto \begin{pmatrix} 0 \\ \sin\theta_0 \\ -1 \\ \cos\theta_0 \end{pmatrix}; \quad |4\rangle \propto \begin{pmatrix} b_4 \\ -\cos\theta_0 \\ 0 \\ \sin\theta_0 \end{pmatrix},$$

$$b_2 = \frac{\sqrt{(3\mathscr{B}^2-m^2)^2+4\mathscr{B}^2k^2}-(3\mathscr{B}^2-m^2)}{2\mathscr{B}\omega_2},$$

$$b_4 = \frac{-\sqrt{(3\mathscr{B}^2-m^2)^2+4\mathscr{B}^2k^2}-(3\mathscr{B}^2-m^2)}{2\mathscr{B}\omega_4}. \tag{G.9}$$

At small momenta, $k \to 0$, the coefficients behave as $b_2 \to 0, b_4 \to \text{const} \neq 0$. Note that there is no naive orthogonality of the eigenvectors, in particular $\langle 2|4\rangle \neq 0$. A similar nonorthogonality is typical for classical (non-quantum) coupled oscillators in magnetic field. In spite of the nonorthogonality, after an appropriate canonical transformation the Hamiltonian is transformed to the sum of independent Hamiltonians of the normal modes.

G.2 Spinor Excitations in Vector Representation

Using Eqs. (G.1) and (G.4) we find,

$$z_+ \propto \begin{pmatrix} [(a_+ + a_-)\cos\frac{\theta_0}{2} - (b_+ + b_-)\sin\frac{\theta_0}{2}] \\ [(a_+ + a_-)\sin\frac{\theta_0}{2} + (b_+ + b_-)\cos\frac{\theta_0}{2}]e^{i\varphi_0} \end{pmatrix} \tag{G.10}$$

$$z_- \propto \begin{pmatrix} [(a_+ - a_-)\cos\frac{\theta_0}{2} - (b_+ - b_-)\sin\frac{\theta_0}{2}] \\ [(a_+ - a_-)\sin\frac{\theta_0}{2} + (b_+ - b_-)\cos\frac{\theta_0}{2}]e^{i\varphi_0} \end{pmatrix} \tag{G.11}$$

From here, using

$$\delta\vec{\zeta} = \delta z^\dagger \vec{\sigma} z_0 + z_0^\dagger \vec{\sigma}\delta z, \tag{G.12}$$

together with Eqs. (G.2) and (G.9) we find explicit formulas for the spin vibration vectors $\delta\vec{\zeta}$ presented in Eqs. (9.9) and (9.10).

References

1. Rüegg C, Normand B, Matsumoto M, Furrer A, McMorrow DF, Krämer KW, Güdel HU, Gvasaliya SN, Mutka H, Boehm M (2008) Quantum magnets under pressure: controlling elementary excitations in $TlCuCl_3$. Phys Rev Lett 100:205701
2. Rüegg C, Furrer A, Sheptyakov D, Strässle T, Krämer KW, Güdel H-U, Mélési L (2004) Pressure-induced quantum phase transition in the spin-liquid $TlCuCl_3$. Phys Rev Lett 93:257201
3. Zinn-Justin J (2002) Quantum field theory and critical phenomena. Clarendon Press, International series of monographs on physics
4. O'Connor D, Stephens CR (1991) Phase transitions and dimensional reduction. Nuclear Phys B 360(2):297–336
5. O'Connor D, Stephens CR (1994) Crossover scaling: a renormalization group approach. Proc R Soc Lond A Math Phys Eng Sci 444(1921):287–296
6. Freire F, O'Connor D, Stephens CR (1994) Dimensional crossover and finite-size scaling below T_c. J Stat Phys 74(1):219–238
7. O'Connor D, Stephens CR (1994) Effective critical exponents for dimensional crossover and quantum systems from an environmentally friendly renormalization group. Phys Rev Lett 72:506–509
8. O'Connor D, Stephens CR (1994) Environmentally friendly renormalization. Int J Mod Phys A 09(16):2805–2902
9. O'Connor D, Stephens CR (2002) Renormalization group theory of crossovers. Phys Rep 363(4–6):425–545
10. Sachdev S (1997) Theory of finite-temperature crossovers near quantum critical points close to, or above, their upper-critical dimension. Phys Rev B 55:142–163
11. Pisarski RD (1988) Computing finite-temperature loops with ease. Nuclear Phys B 309(3):476–492

Printed in the United States
By Bookmasters